Glossary of Supply Chain Terminology

For Logistics, Manufacturing, Warehousing, and Technology

Philip Obal

Industrial Data & Information Inc. (IDII)

www.IDII.com

© 2004 Industrial Data & Information, Inc.

**GLOSSARY OF SUPPLY CHAIN TERMINOLOGY
FOR LOGISTICS, MANUFACTURING, WAREHOUSING,
AND TECHNOLOGY**

BY PHILIP OBAL

Copyright © 2004 by Industrial Data & Information Inc. All rights reserved. Except as permitted under the Copyright Act of 1976, no part of this publication may be reproduced or distributed in any form or by any means, or stored in a database or retrieval system, without the prior written permission of the publisher, Industrial Data & Information, Inc.

International Standard Book Number: 0-9669345-3-9
Library of Congress Card Number: 2003112569
Printed in United States of America

Printing History:
    September 2003   First Edition
    January 2004   Second Edition

All company and product names are the trademarks, registered trademarks, or marks of their respective companies or organizations.

Quotes must have written permission from the publisher as indicated above. Quotes from this publication must also include (1) the full name of this publication *"Glossary of Supply Chain Terminology - For Logistics, Manufacturing, Warehousing, And Technology – 2nd Edition – January 2004* and (2) our website URL of **www.IDII.com.**

Industrial Data & Information, Inc.
Route 1, Box 580
Webbers Falls, OK 74470
U.S.A.

918-464-2222 Voice
918-464-2221 Fax

www.IDII.com

Glossary of Supply Chain Terminology

Acknowledgments

I thank God for my excellent wife, Melinda Obal, of 23 years, who has been a tremendous inspiration & encouragement to me. We also thank Kristina LaBounty & Elizabeth Obal for the many hours of working on this glossary.

In addition to our definitions of words & acronyms, the glossary now includes contributions of fellow professionals. We have reviewed & included many of their **professional & outstanding** definitions into our glossary. The following individuals & companies deserve special recognition:

Our Appreciation To These Additional Contributors

Ken Ackerman (KA), http://www.warehousingforum.com, for permission to use his definitions. Ken is an expert warehousing & management consultant, with an MBA from Harvard. He has authored many books for warehousing & operations. See **"Warehousing Forum"** newsletter and Ken's latest book **"Auditing Warehouse Performance"**.

William J. Augello (WA), http://www.transportlawtexts.com, for permission to use his glossary on legal & transportation terms. Bill is Counsel to the law firm of Auguello, Pezold, and Hirschmann P.C., and teaches transportation law at University of Arizona. Bill has written many legal briefs, articles, and books, including the popular **"Freight Claims in Plain English"** and recent **"Transportation, Logistics, and the Law"**.

Dr. Edward J. Marien (EM) of the University of Wisconsin-Madison, for use of their acronym list. Dr. Edward Marien is the Professor and Program Director of SCM Executive Education Programs. He holds degrees from Indiana University (MBA) and Michigan State University (Ph.D., Marketing and Logistics). (608) 441-7304.

John Seidl, http://www.tompkinsinc.com, for permission to use his definitions. John is a Partner with Tompkins Associates where he is responsible for **their supply chain technology systems**

Suggest New Words http://www.IDII.com/addword.htm

© 2004 Industrial Data & Information Inc.

Glossary of Supply Chain Terminology

integration practice. John has authored several articles on supply chain technology and speaks frequently at industry events.

CNF Inc. (Cnf), http://www.cnf.com, for permission to use their glossary on transportation terms. CNF is **transporting freight, managing warehouses, completing final configuration, and distributing the finished goods.** Con-Way and Menlo Logistics are subsidiaries of CNF. (650) 494-2900

FloStor (Fs), http://www.flostor.com , for permission to use their conveyor glossary. FloStor is a leading west coast USA integrator for **design engineering and turnkey installation of automated material handling systems.** Their projects have appeared as eight feature cover articles in *Modern Materials Handling* magazine. (800) 500-8256

Hach Company (HC), http://www.hach.com, for permission to use their MSDS words & definitions. Hach manufactures and distributes **analytical instruments, testing equipment, and reagents** to test water and aqueous solutions

Pallet One (P1), http://www.palletone.com, for permission to use their pallet words & definitions. **Pallet One is the largest new pallet manufacturer** in the USA and operates 12 facilities and employs over 1,000 people. Pallet One provides pallets, pallet-related products, and wood-based by-products. (800) 771.1148.

To give proper credit & recognition, words from the above-contributed sources will have their initials following the definition.

We also recognize & thank the **many individuals** who have suggested new words & definitions to the glossary!

Future Contributors

If you have developed a dictionary or glossary, please email us the glossary with a letter stating that you are granting Industrial Data & Information Inc. permission to use the information in our publications. **We *greatly* appreciate your help!**

Suggest New Words   http://www.IDII.com/addword.htm

© 2004 Industrial Data & Information Inc.

Glossary of Supply Chain Terminology

Target Audience

This glossary can educate supply chain professionals involved in Information Technology, Logistics, Business, Transportation, Operations, or Warehousing.

The divisions in business, computers, government, manufacturing, transportation, and warehouse operations have their own set of terminology. *Our aim to help these supply chain professionals understand the terminology in this ongoing multi-sector and multi-division environment!*

The following personnel in an organization can benefit from this publication: Operation Management, Warehouse Management, Purchasing Management, Logistics Management, Customer Service Management, Information Technology, and owners of the company who are involved in the details of warehouse control. Consultants, IT Management, and software developers also can benefit from this publication. Since the supply chain software field is gaining interest, Venture Capitalist and other professionals that invest in companies, need this publication to fully understand industry terminology.

Contact the Author

Philip Obal is available for consulting, project management, software design, and speaking engagements. Phil welcomes comments and suggestions on warehousing and supply chain issues. Send these to philobal@idii.com. Our website has our latest phone numbers if one needs to call or send a fax.

Suggest New Words http://www.IDII.com/addword.htm

© 2004 Industrial Data & Information Inc.

Glossary of Supply Chain Terminology

Products from IDII

The *IDII Software Newsletter* provides an education and condensed news regarding supply chain software and delivered twice a month via e-mail. It is for busy professionals that are involved in software for the warehouse, logistics, and the supply chain. See www.idii.com/esn/index.htm to register to receive the newsletter free of charge.

The *Supply Chain Events Newsletter* provides a detailed list of educational seminars, classes, Webinars, conferences, and trade shows in the supply chain. This newsletter is sent out twice a month via e-mail. See www.idii.com/esn/index.htm to register to receive the newsletter free of charge.

IDII provides **WMS RFP's & TMS RFP's** for those seeking a new Warehouse Management System (WMS) or Transportation Management System (TMS). Our advanced WMS & TMS RFP products will **save significant time** and enables one to be **very thorough** during the software selection process.

IDII produces a *"Warehouse & Logistics Software Directory"* which is an **exhaustive** listing of software solutions for warehouse operations. It is updated yearly, since software solutions are continually upgraded.

See our educational research report on *"What To Look For In Warehouse Management System Software"*. This gives many **insights on common differences & deficiencies** in warehouse software – which raises the knowledge level of the selection team members, consultants, and IT staff.

New research on WMS solutions is also available. IDII is working on **"Leading WMS Solutions"** research report currently and this report is due out in January 2004.

Suggest New Words http://www.IDII.com/addword.htm

© 2004 Industrial Data & Information Inc.

Glossary of Supply Chain Terminology

Table of Contents

- Acknowledgments 1
- Target Audience 3
- Contact the Author 3
- Table of Contents 5
- Word Types 6
- Suggest a New Word 7
- *Business* *8*
- *Computer* *34*
- *Conveyor* *59*
- *EDI* *76*
- *Government* *83*
- *Manufacturing* *87*
- *Material Safety Data Sheets* *95*
- *Organization* *110*
- *Pallet* *129*
- *Purchasing* *140*
- *Standards* *144*
- *Transportation* *152*
- *Warehouse* *197*
- *Main Glossary* *237*
- **Need More Information?** 442
- **Suggest Words, Acronyms, Abbreviations** 442

Suggest New Words http://www.IDII.com/addword.htm

© 2004 Industrial Data & Information Inc.

Glossary of Supply Chain Terminology

Word Types

Word types in the glossary are to assist the individual in understanding the word, acronym, or abbreviation. The following categories or types exist:

Business (B)	Terms & words used by the business community.
Computer (C)	Terms & words utilized by computer professionals.
Conveyor (Conv)	Terminology & abbreviations describing conveyors.
Electronic Data Interchange (EDI)	Established terminology utilized by EDI specialists.
Government (Gov)	Terminology used frequently by the US & other governments.
Manufacturing (Mfg)	Terms & words found in manufacturing.
Material Safety Data Sheets (MSDS)	Terminology utilized with Material Safety Data Sheets
Organization (Org)	Non-profit and for profit organizations.
Pallet (Pa)	Terminology utilized with pallets and containers that hold products.
Purchasing (Pu)	Terms & words found in purchasing & procurement of goods.
Standards (Std)	Established and proposed standards.
Transportation (Tr)	Terms & words utilized in the transportation industry.
Unit of Measure (Uom)	Common unit of measurements and metrics utilized in all the aspects of the supply chain.
Warehouse (W)	Terms & words used in the warehousing industry.

Suggest New Words   http://www.IDII.com/addword.htm

© 2004 Industrial Data & Information Inc.

Glossary of Supply Chain Terminology

Word types are utilized on each defined word. E.G., Pick-To-Carton has word type W (a Warehouse word type). It is just a further refinement to section the glossary into highly usable sections and to properly categorize the word for ease of learning.

Suggest a New Word

Feel free to submit a new term, phrase, or word for consideration into this glossary. We will review it. Enter the new word & definition at http://www.IDII.com/AddWord.htm.

If you have a large list of words, then e-mail this list to editor@idii.com. If you e-mail us a list, please make sure you also grant IDII permission to use these words & definitions in all of our publications & materials.

Business

Terms & Words
In The
Business Community

Glossary of Supply Chain Terminology

Word	Type	Definition of Word
A2A	B C	Application to Application. Connecting software application to another software application. See also EAI, B2B, XML, and EDI.
ABC	B	Activity Based Costing.
Account	B	1) An individual, institution, or other organization that purchases a company's products. 2) The general category of service as listed on the company books. (KA)
Act of God	B	An unavoidable occurrence or accident produced by physical cause such as floods, earthquakes, most fires, or other natural disasters. (KA)
ADA	B	Americans with Disabilities Act. A 1990 US federal law that requires public facilities to be accessible to persons with disabilities.
ADDR	B	Address. Common abbreviation used in business and in computer programming.
Ad Hoc	B	Ad Hoc Queries – An "on the spot" database query that is created by a user.
ADJ	B	Adjustment. Common abbreviation used in business and in computer programming.
AKA	B	Also Known As.
Algorithm	B	A method or process, such as an equation or computation, for solving a problem. (KA)
AMP	B	Advertising, Marketing and Promotion (EM)
AMT	B	Amount. Common abbreviation used in business and in computer programming.
AP	B	Accounts Payable. (1) The departments of a company responsible for verifying & payment of services rendered and products bought from vendors. (2) A software module utilized for entry of payable bills, aging of accounts payable, and printing of checks.

Suggest New Words http://www.IDII.com/addword.htm

© 2004 Industrial Data & Information Inc.

Glossary of Supply Chain Terminology

Word	Type	Definition of Word
AR	B	Accounts Receivable. (1) The department of a company responsible for money due for services performed or merchandise sold on credit. (2) A software module utilized for invoicing, cash application, and aging of accounts receivables.
ATP	B	See Available to Promise.
Audit Trail	B	The records and management controls that document business activities. Receipt, handling, and movement of materials throughout a warehouse are part of an audit trail. (KA)
Available to Promise	B	Current inventory that is available & free from any commitments. This is the current on-hand inventory minus committed inventory for sales orders, transfer orders, Manufacturing orders (when product can also be a component), and vendor returns.
AVG	B	Average. Common abbreviation used in business and in computer programming.
AYPI	B	And Your Point Is? AYPI is an acronym used in messaging other online users.
B2B	B	Business-to-Business.
B2B2C	B	Business to Business to Consumer (EM)
B2C	B	Business-to-Consumer.
B2D	B	Business-to-Distributor.
B2E	B	Business-to-Employee.
B2G	B	Business-to-Government.
Back Order	B	Items that have been ordered but cannot be shipped due to stockout. Merchandise on back order is scheduled for shipment when it becomes available. (KA)
BAL	B	Balance. Common abbreviation used in business and in computer programming.
BFN	B	Bye For Now. BFN is an acronym used in messaging other online users.

Suggest New Words   http://www.IDII.com/addword.htm

© 2004 Industrial Data & Information Inc.

Glossary of Supply Chain Terminology

Word	Type	Definition of Word
BI	B	Business Intelligence. From a computer savvy person, business intelligence is mining data from database(s). For a marketing & sales viewpoint, business intelligence may be both competitive information gathering as well as internal information gathering. Other terms similar to this are BPM, EPM, EIS, and OLAP.
Boilerplate	B	An agreement, purchase order, or other document that includes standard terms and procedures. (KA)
Bottleneck	B W	Congestion or significant slow down at an area due to an inefficient resource or process.
BP	B	Best Practice (EM)
BP	B	Business Process
BPI	B	Business Process Improvement.
BPM	B	Business Performance Management. Other terms similar to this are BI, EPM, EIS, and OLAP.
BPO	B	Business Process Optimization. AKA process re-engineering.
BPO	B	Business Process Outsourcing.
BPR	B	Business Process Reengineering.
BTS	B	Balance To Ship. The remaining quantity of an order to be shipped.
CALC	B	Calculation. Common abbreviation used in business and in computer programming.
CAPA	B	Corrective And Preventive Action. A plan of actions to correct & resolve a problem.
CAR	B	Capital Appropriation Request.
Carrying Costs	B	The cost of holding inventory, including taxes, depreciation, cost of invested capital, and insurance. Expressed as a percentage of total inventory, carrying cost is used in calculating economic order quantities. (KA)
C-Commerce	B	Collaborative Commerce.

Suggest New Words http://www.IDII.com/addword.htm

© 2004 Industrial Data & Information Inc.

Glossary of Supply Chain Terminology

Word	Type	Definition of Word
CEO	B	Chief Executive Officer. Principal individual responsible for the activities of the company.
CFAR	B	Collaboration Forecasting and Replenishment. See also CPFR.
CFO	B	Chief Financial Officer. Principal individual responsible for handling funds, financial planning of the company, and financial records.
CFPIM	B Mfg	Certified Fellow in Production and Inventory Management. Advanced certification program for Manufacturing and inventory by APICS.
CGP	B	Cost of Goods Purchased (EM)
Chargeback	B	Form used for recording transactions involving vendor returns. (KA)
CI	B	Continuous Improvement.
C-Inventory	B	Collaborative Inventory Management.
CIO	B	Chief Information Officer. Principal individual responsible for computer activities of the company.
CIRM	B	Certified in Integrated Resource Management. Certification program by APICS.
CIS	B	Customer Information Systems. See CRM
CLO	B	Chief Logistics Officer.
CLP	B	Country List Price
CMs	B	Contract Manufacturers (EM)
CNT	B	Count. Common abbreviation used in business and in computer programming.
CO	B	Company. Common abbreviation used in business and in computer programming.
COGS	B	Cost Of Goods Sold (EM).
Commodity	B	Any article of commerce. Goods, merchandise.
Consigned Stock	B	Finished goods inventories, in the hands of agents or dealers which are still the property of the supplier. (KA)

Suggest New Words http://www.IDII.com/addword.htm

© 2004 Industrial Data & Information Inc.

Glossary of Supply Chain Terminology

Word	Type	Definition of Word
Consignment	B	A transaction in which the title to goods remains with the shipper (the consignor) until the buyer (the consignee) sells the goods. (KA)
COO	B	Chief Operating Officer.
COS	B	Conditions Of Satisfaction. Specific criteria that measures the process or product, for final determination whether it was successful or not.
Cost of Capital	B	The cost to invest or borrow capital, usually expressed in a yearly percentage that is based on such factors as the current interest rate, or against alternative investments such as government securities. (KA)
Cost of Goods Sold	B	A monetary cost per unit of area to measure the basic cost of operating a warehouse. It represents the costs of the physcial space of a facility and the activities that occur within. The cost components include all the costs of operating the warehouse. (KA)
COTS	B Gov	Commercial Off the Shelf.
Counseling	B	Providing feedback to employees on work-related issues such as technical advice, performance reviews, or even personal problems. (KA)
CPI	B	Consumer Price Index.
CPIM	B Mfg	Certified in Production and Inventory Management. Certification program focused on inventory & Manufacturing by APICS.
CPIO	B	Chief Process Improvement Officer (CM)
CPO	B	Chief Privacy Officer.
CPS	B	Collaborative Planning Solutions. Planning software providing a supply chain plan for a week, month, or longer.
Credit Memo	B	A form used for recording and processing transactions involving a credit to suppliers or buyers. (KA)

Suggest New Words http://www.IDII.com/addword.htm

© 2004 Industrial Data & Information Inc.

Glossary of Supply Chain Terminology

Word	Type	Definition of Word
CRO	B	Chief Risk Officer. CROs are on either the board or an executive in the company to manage all risk within the company, whether it is operational, information technology, credit, market, or reputation risk. Some IT managers prefer to be reporting to the CRO rather than the CFO.
CS	B	Customer Service.
CSO	B	Chief Security Officer.
CSR	B	Customer Service Representative. A person with heavy contact with the customer. May be dedicated to one or more customers and duties will vary dependent upon type of industry (3PL, Manufacturer, Distributor).
CTD	B	Capable To Deliver. Order commitment based upon available inventory quantities and pipeline (manufacturing or in-transit quantities).
CTO	B	Chief Technology Officer.
CTP	B	Capable to Promise.
CTQ	B	Critical to Quality.
CUL8R	B	See You Later. CUL8R is an acronym used in messaging other online users.
CUST	B	Customer. Common abbreviation used in business and in computer programming.
Customer Service	B	The aspect of logistics which provides the highest level of service at the lowest cost. An all-encompassing term for non-price and non-physical product features of a product, usually consisting of service lead time, customer contact and problem resolution. (KA)
DAM	B	Digital Asset Management (EM)
DE	B	Debt/Equity ratio (EM)

Suggest New Words http://www.IDII.com/addword.htm

© 2004 Industrial Data & Information Inc.

Glossary of Supply Chain Terminology

Word	Type	Definition of Word
Delivered Price	B	A price for merchandise that includes transportation charges to a delivery point agreed upon by the seller and buyer. (KA)
Deregulation	B	The reduction of governmental control of business, especially to permit increased competition in a free market. (WA)
Discrimination	B	The effect of a statute or established practice that creates disadvantages for a certain group of persons or interests. (WA)
Distribution System	B	The system and processes of transporting goods within and among plants, warehouses, and other facilities. (KA)
Distributor	B	A business that is in the middle of a supply channel. Distributors buy and sell finished goods. They man alter, assemble, combine, or otherwise add value to the goods. (KA)
Diversification Strategy	B	Spreading a company's risk in case of a downturn in the demand for any given product or service. Diversification may involve expansion into different markets or product lines. (KA)
DL	B	Direct Labor
DMAIC	B	Define, Measure, Analyze, Improve, Control.
Downtime	B	The time when equipment is scheduled for operational, but is idle for maintenance, repairs, or changeovers. (KA)
DPM	B Mfg	Defects Per Million.
Drawback	B	Refund of customs duties paid on material imported and later exported. Also known as "Duty Drawback". (Cnf)
DSD	B W	Direct Store Delivery. See also FRM.
DT	B	Date. Common abbreviation used in business and in computer programming.
DUNS	B	Data Universal Numbering System. The Dun & Bradstreet code (DUNS Number) assigned to uniquely identify a company.

Suggest New Words http://www.IDII.com/addword.htm

© 2004 Industrial Data & Information Inc.

Glossary of Supply Chain Terminology

Word	Type	Definition of Word
DuPont Model	B	A display of the factors of return on investment which pinpoints causes for increase or decrease in return. The model divides a business into three areas: financial management, asset management, and operations. Pretax return on investment (PROI) equals financial leverage times asset turnover times net profit margin. Originally developed by E.I. du Pont de Nemours. (KA)
Duty	B	A taxed assessed by the government for importing and exporting goods. (KA)
Duty Drawback	B	Refund of customs duties paid on material imported and later exported. Also known as Drawback. (Cnf).
EAS	B	Electronic Article Surveillance.
E-business, EB	B	Electronic Business.
EC, E-commerce	B	Electronic Commerce.
ECR	B W	Efficient Customer Response.
EFF	B	Effective. Common abbreviation used in business and in computer programming.
EFQM	B	European Foundation for Quality Management (EFQM) Excellence Model.
EFT	B	Electronic Funds Transfer.
EHCR	B W	Efficient Healthcare Consumer Response. To realize supply chain cost savings through the adoption of EDI, bar-coding, and other strategies such as Activity Based Costing. See also ECR.
EIN	B	Employer Identification Number
EMEA	B	Europe, Middle East, Africa.
Enterprise Resource Planning	B	The corporate computer systems with accounting, order management, purchasing, and inventory software. Also termed the "Host System", when one is integrating a WMS or TMS with the ERP solution.

Suggest New Words http://www.IDII.com/addword.htm

© 2004 Industrial Data & Information Inc.

Glossary of Supply Chain Terminology

Word	Type	Definition of Word
EPS	B	Earnings Per Share.
ERM	B	Employee Relationship Management. Software to assist in employee information & management thereof. CRM vendor Siebel (Nasdaq: SEBL) has produced an ERM software module in early 2003.
ERM	B	Enterprise Relationship Management. See also CRM, PRM, and SRM.
ERP	B	See Enterprise Resource Planning.
ERP II	B	Enterprise Resource Planning II. Some view Advanced ERP as ERP. A matter of semantics. Common term is still ERP.
EST	B	Estimate or Estimated. Common abbreviation used in business and in computer programming.
EULA	B C	End User License Agreement. A contractual agreement for usage of a software package or operating system.
Excess Stock	B	Any quantity of inventory, either held or on order, which exceeds known or anticipated forward demand to such a degree that disposal action should be considered.
Exempt Positions	B	Positions as defined by the Fair Labor Standards Act that do not require overtime payments since they are executive, supervisory, or administrative. (KA)
Exit Interview	B	An interview with an employee who is leaving a company. The purpose is to discover the reasons for the individual's departure. (KA)
FAQ	B	Frequently Asked Questions. Detailed answers to frequently asked questions. Designed to help the customer 24x7 days a week. Many companies on their websites have FAQ pages.

Suggest New Words http://www.IDII.com/addword.htm

© 2004 Industrial Data & Information Inc.

Glossary of Supply Chain Terminology

Word	Type	Definition of Word
FAT	B	Factory Acceptance Test. The client company is testing the stand-alone performance & functionality that is built-into the software solution.
FCST	B	Forecast.
Fill Rate	B	A measurement of how well a warehouse is meeting service objectives. It is calculated by dividing the number of orders filled by the total number of orders within a given period. Also Order Ratio. (KA)
FIM	B	Fresh Item Management (EM)
FIT	B Tr	Fabrication in Transit.
Foreign Trade Zones	B Tr	Goods subject to duty may be brought into such zones duty-free for transshipment/storage/minor manipulation/sorting. Duty must be paid when/if goods are brought from a zone into any part of the U.S. (Cnf)
FRU	B	Field Replaceable Unit
FS	B	Field Service
FTD	B	Foreign Trade Division
FTE	B	Full Time Employees.
FTE	B	Full Time Equivalent.
FTSR	B	Foreign Trade Statistical Regulations
FTZ	B Tr	See Foreign Trade Zone.
GAAP	B	Generally Accepted Accounting Principles.
GOWI	B	Get On With It. GOWI is an acronym used in messaging other online users.
GWTP	B	Get With The Program. GWTP is acronym used in messaging other online users.
HCM	B	Human Capital Management.

Suggest New Words http://www.IDII.com/addword.htm

© 2004 Industrial Data & Information Inc.

Glossary of Supply Chain Terminology

Word	Type	Definition of Word
Host System	B	See Enterprise Resource Planning. A "host system" refers to the current enterprise software residing on 1 or more computers. When planning for additional software, one must interface to the host system. When planning for new enterprise wide software, one must convert data from the current host system into the new software.
HQ	B	Headquarters.
HR	B	Human Resources. Either a department within the company dealing with management of employees or a software module that this department heavily utilizes.
HTH	B	Hope This Helps. HTH is an acronym used in messaging other online users.
IAW	B	In Accordance With.
ID	B	Identification number or unique identification code value. Common abbreviation used in business and in computer programming.
IL	B	Indirect Labor.
IMHO	B	In My Humble Opinion. IMHO is an acronym used in messaging other online users.
Indirect Costs	B	Costs that can not be directly associated with specific goods or services such as utilities, marketing, and staff functions. These are typically allocated to a final product through an overhead account. (KA)
Interstate Commerce	B	Exchanging goods between buyers and sellers in two or more states. (Cnf)
Intrastate Commerce	B	When all business between buyers and sellers is carried on within state. (Cnf)
INV	B	Inventory.
Inventory Control	B	The activities and techniques associated with maintaining the optimal level and location of raw materials, work-in-progress, and finished goods in a supply chain. (KA)

Suggest New Words http://www.IDII.com/addword.htm

© 2004 Industrial Data & Information Inc.

Glossary of Supply Chain Terminology

Word	Type	Definition of Word
Invoice	B	A sales document evidencing the points of purchase, such as quantity, price, weight, etc. (WA)
IP	B	Intellectual Property (EM)
IPO	B	Initial Public Offering.
IRR	B	Internal Rate of Return.
ITO	B	Information Technology Outsourcing.
KPI	B	Key Performance Indicator. Metrics.
KSF	B	Key Success Factor.
Landed Cost	B	Total expense of receiving goods at place of retail sale, including retail purchase price and transportation charges. (Cnf)
Lead Time	B	The period of time that elapses between the time an order is placed and the time it is received in storage. This is often called the replenishment time. (KA)
Lead Time Inventories	B	A quantity that is sufficient to meet demand between the time a replenishment is ordered until it is delivered. See Reorder Point Quantity. (KA)
Learning Curve	B	A term used to describe the ability of individuals to become better at their job and more productive as they gain experience. (KA)
LN	B	Line. Common abbreviation used in business and in computer programming.
LOB	B	Line of Business.
LOC	B	Location. Common abbreviation used in business and in computer programming.
LOI	B Tr	Letter of Intent.
LOL	B	Laughing Out Loud. LOL is an acronym used in messaging other online users.
LOV	B	List of Values (EM)
LSP	B	Logistics Service Provider.

Suggest New Words http://www.IDII.com/addword.htm

© 2004 Industrial Data & Information Inc.

Glossary of Supply Chain Terminology

Word	Type	Definition of Word
MAD	B	Mean Absolute Deviation.
Maintenance, Repair, and Operating Items and Supplies	B	Abbreviated MRO. All items used maintaining, repairing and operating a facility. Maintenance items are those necessary to keep the facility in optimum operating condition, such as janitorial supplies. Repair items are those necessary to keep the facility and machinery functioning, such as replacement parts. Operating items are those required for the operation such as fuel and office supplies. (KA)
Management by Objectives	B	Abbreviated MBO. A participative method of motivational management. Goals are determined with input from subordinates. (KA)
MAX	B	Maximum. Common abbreviation used in business and in computer programming.
MBO	B	Management Buy Out.
Merchandise	B	Goods bought and sold in trade, with the exception of real estate, cash, and negotiable paper. (KA)
MICR	B C	Magnetic Ink Character Recognition. Special ink with magnetic abilities to be read by high peed readers. Commonly used in personal & business checks.
MIN	B	Minimum. Common abbreviation used in business and in computer programming.
MRO	B	See Maintenance, Repair, and Operating Items and Supplies. (KA)
MSG	B	Message. Common abbreviation used in business and in computer programming.
MTD	B	Month To Date. Showing a summary, sub-total, or total value that is a month to date value.

Suggest New Words http://www.IDII.com/addword.htm

© 2004 Industrial Data & Information Inc.

Glossary of Supply Chain Terminology

Word	Type	Definition of Word
NAFTA	B	North American Free Trade Agreement. NAFTA is a free trade agreement between Canada, the United States, and Mexico to eliminate trade related tariffs and duties, as well facilitate increased environmental, transportation, and labor-related safety. (WA)
NFR	B	Not For Resale. An indicator by the manufacturer that this product is not intended to be resold. This version of the product is either free or heavily discounted, while the same product normally sells higher in its regular price range.
Noncontiguous	B	A possession of a sovereign nation although not geographically interconnected. (WA)
Non-Exempt Positions	B	Employees who are not assigned to supervisory or executive positions. These workers are eligible for overtime. (KA)
NPD	B	New Product Development.
NVP	B	Net Present Value.
OBI	B	Open Buying on the Internet. A project of the OBI Consortium (OBIC) in 1996 by major Fortune 500 companies in the US.
Obsolescence	B	The cost associated with inventory becoming obsolete. It can be extended to include such costs as disposal and storage. (KA)
OE	B	Operational Excellence.
OEE	B Mfg	Operating Equipment Effectiveness. Business metric to measure the production line.
OEM	B	Original Equipment Manufacturer.
OJT	B	On-the Job Training.
OMB	B	Office of Management and Budget
OOS	B	Out Of Specification. When a sample of the product has been tested by QA/QC and found to be out-of-specification.

Suggest New Words http://www.IDII.com/addword.htm

© 2004 Industrial Data & Information Inc.

Glossary of Supply Chain Terminology

Word	Type	Definition of Word
Operating Efficiency	B	The percentage of actual output compared to standard output. (KA)
Opportunity Costs	B	The expected returns of one activity which are foregone in order to pursue other activities or projects. See Alternative Cost. (KA)
Optimization	B	Achieving the best possible solution or results with given resources and constraints. (KA)
OR	B	Operations Research.
Ordering Costs	B	All of the costs associated with the task of preparing, transmitting, following up, and recording the receipt of an order. (KA)
OTS	B	Off The Shelf.
Outsourcing	B	Using a third-party manufacturer, carrier, or warehouse to perform functions formerly assigned to employees. (KA)
Overhead Allocation	B	The portion of overhead cost that is applied by accountants to a particular product or service. (KA)
Pareto's Law	B	A rule that says a relatively small number of products, sales, or activities compromise a large percentage of the total. First described by Italian sociologist Vilfredo Pareto. Also called the 80/20 rule. See ABC Management. (KA)
PARM	B	Parameter. Common abbreviation used in business and in computer programming.
PCT	B	Percent. Common abbreviation used in business and in computer programming.
PERT	B	Program and Evaluation Review Technique.
Physical Distribution	B	A term that applies to the logistics activities that occur between the end of the production line and the final user. It includes traffic, packaging, materials handling, warehousing, order entry, customer service, inventory control, and forecasting. (KA)

Suggest New Words http://www.IDII.com/addword.htm

© 2004 Industrial Data & Information Inc.

Glossary of Supply Chain Terminology

Word	Type	Definition of Word
Pilferage	B	1) The felonious act of breaking into containers and removing property. 2) Petty theft of merchandise. (KA)
Pipeline Stock	B	Inventory within the pipeline, including in-transit inventory as well as inventory positioned in distribution centers. (KA)
P&L	B	Profit and Loss.
PLC	B	Public Limited Company.
PLM	B	Product Lifecycle Management. A way to manage product data from the concept stage until a product sunset's years later. Software vendors are promoting this valuable concept.
PLU	B	Price Look Up.
PM	B	Preventive Maintenance. Equipment needs scheduled PM for cleaning and replacement of maintenance components.
PN, P/N	B	Part Number. A code to uniquely identify a part/product. See UPC.
PO	B Pu	Purchase Order. Document sent to a vendor indicating goods or services to be purchased at a specific price and terms for indicated quantities. For EDI see 850.
POI	B	Point Of Installation (EM)
POS	B	Point of Sale.
Postponement	B	The delay of the final packaging, branding, formulation, or commitment of a product until the last possible moment. The practice reduces speculation or risk by delaying differentiation of the product until it is purchased. (KA)
POU	B	Point Of Use (EM)
Preventive Maintenance	B	Actions taken to reduce break-downs in machinery and equipment, such as basic cleanliness and replacement or adjustment of parts. (KA)
PRM	B	Partnership Relationship Management. See also CRM, ERM, and SRM.

Suggest New Words   http://www.IDII.com/addword.htm

© 2004 Industrial Data & Information Inc.

Glossary of Supply Chain Terminology

Word	Type	Definition of Word
PRM	B	Price-Revenue Management. SCM must take into effect both CRM & PRM issues.
PRO	B	Price Revenue Optimization. A methodology or software that determines which products and services should be price at, in order to maximize profits.
Product Life Cycle	B	The stages a product evolves through during the period that it is marketed. The four general phases are introduction, growth, maturity, and marketing. (KA)
Product Line	B	A group of similar products that are aggregated for planning and marketing. (KA)
Promotional Product	B	A product that is the subject of a campaign to stimulate sales by price reductions or other incentives. This process may cause a wide fluctuation in demand. AKA Promo. (KA)
PSA	B	Product Service Agreements (EM)
PTD	B	Period-To-Date. Common abbreviation used in business and in computer programming.
PTD	B	Profitable To Deliver. Order commitment based upon CTD and the profitability that includes the execution side (transportation, distribution, logistics). See also PTP.
PTP	B	Profitable To Promise. In addition to ATP and CTP, the PTP analyzes profitability of all factors involved in this order fulfillment.
Pull Distribution	B	System in which retail demand stimulates inventory and transportation flows. (KA)
Purchase Order	B	Document sent to a vendor indicating goods or services to be purchased at a specific price and terms for indicated quantities. Abbreviated PO. For EDI see 850.
Push Distribution	B	A system in which distribution centers and retail points are stocked in anticipation of demand. (KA)

Suggest New Words http://www.IDII.com/addword.htm

© 2004 Industrial Data & Information Inc.

Glossary of Supply Chain Terminology

Word	Type	Definition of Word
QoS	B C	Quality of Service. Describes fast network speeds with quality of service option
QR	B	Quick Response.
QRP	B	Quick Response Program.
QTD	B	Quarter-To-Date. Common abbreviation used in business and in computer programming.
QTY	B	Quantity. Common abbreviation used in business and in computer programming.
Quality Circle	B	A team of workers that meets to solve quality control problems. (KA)
Quantity Discount	B	A reduction in the purchase price that results from increasing of the quantity or value of the overall order. (KA)
Queue	B	The line formed by loads, items or people while waiting. (KA)
R2R	B	Reseller-to-Reseller. Similar to B2B, but each party involved is a reseller. Therefore, the product or service being sold by the reseller (R2R) is purchased by another reseller for resale.
RCPT	B	Receipt. Common abbreviation used in business and in computer programming.
R&D	B	Research & Development.
REC	B	Record. Common abbreviation used in business and in computer programming. Usually refers to a specific record in a database table.
REF	B	Reference. Common abbreviation used in business and in computer programming.
Replacement Cost	B	The current fair-market price to purchase an equivalent asset. (KA)
Reserved Inventory	B	Inventory that is held in reserve for one or more customers.(KA)

Suggest New Words http://www.IDII.com/addword.htm

© 2004 Industrial Data & Information Inc.

Glossary of Supply Chain Terminology

Word	Type	Definition of Word
Reverse Logistics	B W	The requirement to plan the flow of surplus or unwanted material or equipment back through the supply chain after meeting customer demand.
RFI	B	Request For Information. A request for a vendor to provide information. Not as intensive as a RFP, but may provide detailed functional answers to what a solution can provide.
RFP	B	Request For Proposal. A request for a vendor to do a full proposal including costs, references, project plans (implementation), and detailed functional answers from a vendor to a prospective client.
RFQ	B	Request For Quotation. A cost proposal for a given service or product. High-level summaries are given on functions.
RM	B Tr	Revenue Management. Determination of total cost, labor, and how to determine pricing for revenue analysis. Being done especially with Cargo Revenue Management.
RMA	B	Return Merchandise Authorization.
ROA	B	Return on Assets.
ROE	B	Return on Equity.
ROI	B	Return On Investment. Calculations that result in a measurement of what a company will gain from an investment.
ROO	B	Return On Opportunity. Calculations that result in a measurement of what a company will gain from an opportunity.
RPM	B	Real-Time Performance Management. Business metrics to measure performance in real-time.
RPT	B	Report. Common abbreviation used in business and in computer programming.
RSN	B	Reason. Common abbreviation used in business and in computer programming.

Suggest New Words http://www.IDII.com/addword.htm

© 2004 Industrial Data & Information Inc.

Glossary of Supply Chain Terminology

Word	Type	Definition of Word
Salable Goods	B	Products authorized for sale to customers. (KA)
SAT	B	Site Acceptance Test. See also Factory Acceptance Test (FAT).
SC	B	Supply Chain.
SCM	B	See Supply Chain Management.
SCO	B	Supply Chain Optimization.
SCOR	B	Supply Chain Operations Reference-Model.
SCP	B	Supply Chain Planning.
SCV	B	Supply Chain Visibility.
SDR	B	See Special Drawing Rights.
SEQ	B	Sequence. A sequential number. Common abbreviation used in business and in computer programming.
SITD	B	Still In The Dark. SITD is an acronym used in messaging other online users.
Six Sigma	B	Quality management model.
SLA	B C	Service Level Agreement. A written contract where servicing company agrees to adhere to provide service(s) that are at or above a specified service level.
SLM	B	Service Lifecycle Management (EM)
SMB	B	Small to Medium Business. The market segment of businesses that are of small to medium size.
SME	B	Small to Medium Enterprises market. Marketing in software companies will plan target certain size of companies as well as specific vertical markets.
SMM	B Mfg	Small and Medium sized Manufacturers. A grouping of manufacturers by annual revenue size.
SO	B	Sales Order. A customer's order is known as a sales order.
SOA	B	Services-Oriented Architecture (EM)
SOE	B	Statement Of Expectations (EM)

Suggest New Words http://www.IDII.com/addword.htm

© 2004 Industrial Data & Information Inc.

Glossary of Supply Chain Terminology

Word	Type	Definition of Word
SO/HO	B	Small Office/Home Office
S&OP	B	Sales and Operations Planning.
SOP	B	Standard Operating Procedures. Documented procedures, found in wide variety of topics, including GMP.
SPEC	B	Specification. Common abbreviation used in business and in computer programming.
Special Drawing Rights	B	A concept of payment used in international treaties to determine a payment regime. In essence, SDR's are an international currency derived from the average of certain nation's currencies. Abbreviated SDR. (WA)
Standard Costs	B	The anticipated cost for a product or service. (KA)
Statue of Limitations	B	The amount of time allotted under statute before an actionable claim expires and the claimant may no longer enforce her right to be heard. (WA)
STD	B	Standard. Common abbreviation used in business and in computer programming.
STD	B	Standard Deviation
Stock Order	B	An order to replenish depleted inventory, as opposed to an order to fill a customer order or manufacturing requirement. (KA)
Stockout Percentage	B	A customer service measurement that shows a measurement of total stockouts to total orders. (KA)
Stock Report	B	A record of items on hand by type and number based on the paper recording of receipts and shipments during a given period. (KA)
Stock Requisition	B	An intracompany form used to authorize the removal of merchandise from its storage location. (KA)
SUM	B	Summary. Common abbreviation used in business and in computer programming.
Sunset	B	To legislate an act out of existence. (WA)

Suggest New Words http://www.IDII.com/addword.htm

© 2004 Industrial Data & Information Inc.

Glossary of Supply Chain Terminology

Word	Type	Definition of Word
Supply Chain Management	B	Supply Chain Management encompasses the planning and management of all activities involved in sourcing and procurement, conversion, and all Logistics Management activities. Importantly, it also includes coordination and collaboration with channel partners, which can be suppliers, intermediaries, third-party service providers, and customers. In essence, Supply Chain Management integrates supply and demand management within and across companies. Abbreviated SCM. See also SCE and SCP. (Cnf)
TBD	B	To Be Determined.
TCO	B	Total Cost of Ownership. Look at the total cost of ownership for any new project. I.E., new ERP software – the TOC would include hardware, software, professional services, internal staff costs, software installation, and one to three year of maintenance, upgrade, & enhancement costs.
TIN	B	Taxpayer Identification Number. US government requires a TIN on most business transactions for tax reporting purposes.
T&M	B	Time and Materials.
TO	B W	Transfer Order. An internal company order to move inventory from one part of the company to another. This could be stock transfer between warehouses, or warehouse to a van, or division to division.
Tolerance	B	In statistics, the permitted amount of deviation from the mean or average of a measure. (KA)
Tolerated Accuracy Level	B	The percentage that denotes items counted that were within the recorded book value, plus or minus an acceptable limit. (KA)

Suggest New Words http://www.IDII.com/addword.htm

© 2004 Industrial Data & Information Inc.

Glossary of Supply Chain Terminology

Word	Type	Definition of Word
TOT	B	Total. Common abbreviation used in business and in computer programming.
Total Quality Management	B	Abbreviated TQM. An interfunctional approach to process improvement that defines quality by looking at customer expectations. The company then focuses on the prevention, detection, and elimination of defects or other quality problems. (KA)
TPA	B	Trading Partner Agreement. A formal legal document to authorize data exchange between the company and its customers, vendors, carriers, bank, or other trading partner.
TQM	B	Total Quality Management.
Transfer Price	B	The price used when one division of a corporation transfers goods or services to another division. (KA)
TSCM	B	Total Supply Chain Management.
TTM	B	Time To Market. The time it takes a new product to be designed, manufactured, and delivered to the market place. A key metric for how well a company performs in delivering new products.
TTYL	B	Talk To You Later. TTYL is an acronym used in messaging other online users.
Turnover Rate	B	The frequency with which total inventory or a specific class of inventory is completely replaced. It is generally stated as the number of turns per year or per month. (KA)
UM	B	Unified Messaging. UM services combine voice mail, e-mail and fax messages into one mailbox, allowing a user to retrieve and manage all of their messages from a single point, such as a desktop computer or mobile device.
Unreasonable Trade Practice	B	A business practice in violation of the Federal Trade Commission Act. (WA)

Suggest New Words http://www.IDII.com/addword.htm

© 2004 Industrial Data & Information Inc.

Glossary of Supply Chain Terminology

Word	Type	Definition of Word
UR	B	You Are. UR is an acronym used in messaging other online users.
VAD	B	Value Added Distributor. A company that sells product AND offers services (e.g. vendor certifications, technology seminars). A VAD should have the resources to help complete an installation or assist a VAR in doing such.
Variable Costs	B	Costs that vary in proportion to activity levels. (KA)
Variance	B	1) The difference between actual and standard costs, or between budgeted and actual expenditures. 2) In statistics, the measure of dispersion of a distribution. (KA)
VAT	B	Value Added Tax (EM)
VEND	B	Vendor. Common abbreviation used in business and in computer programming. Sometimes abbreviated as VNDR.
Vendor	B	1) A term used interchangeably with supplier. 2) The company that provides materials for production or resale. (KA)
VMI	B	Vendor Managed Inventory
VNDR	B	Vendor. Common abbreviation used in business and in computer programming. Sometimes abbreviated as VEND.
Voucher	B	A document authorizing the disbursement of payment that also signals recognition of a service performed or product purchased. (KA)
WO	B Mfg	Work Order. An order to build finished or semi-finished goods from a bill of materials.
Workers' Compensation	B	Formerly known as workmen's compensation, this is the partial or full replacement of an employee's earnings due to an occupational injury or disease. (KA)

Suggest New Words http://www.IDII.com/addword.htm

© 2004 Industrial Data & Information Inc.

Glossary of Supply Chain Terminology

Word	Type	Definition of Word
WTD	B	Week-To-Date. Common abbreviation used in business and in computer programming.
WW	B	Worldwide
Y2K	B C	Year 2000. Older software kept years in a 2-digit year. Year 2000 compliant software has 4 digit year and leap year issues.
YTD	B	Year To Date. Showing a summary, sub-total, or total value that is for a year to date value.

Suggest New Words http://www.IDII.com/addword.htm

© 2004 Industrial Data & Information Inc.

Glossary of Supply Chain Terminology   34

Computer

Terms & Words
Utilized By
Computer Professionals

Suggest New Words http://www.IDII.com/addword.htm

© 2004 Industrial Data & Information Inc.

Glossary of Supply Chain Terminology 35

Word	Type	Definition of Word
2PC	C	Two Phase Commit
3-Tier	C	See N-Tier.
400	C	Web browser error for "Bad File Request". Check the URL and correct it. Web page or file may not exist.
401	C	Web browser error for "Unauthorized". Failed security authentication. Server requires login and password.
403	C	Web browser error for "Forbidden/Access Denied". Password is required to access this web page, directory, or entire site. See also 401.
404	C	Web browser error for "File not found". Could not find web page. Check typing of the URL string and retry. Site could be down or web page no longer exists.
408	C	Web browser error for "Request Timeout". Request not filled due to time limit exceeded or user canceled the request.
500	C	Web browser error for "Internal Error". Server problem, contact webmaster.
501	C	Web browser error for "Not Implemented". Server does not support the feature requested.
502	C	Web browser error for "Service Temporarily Overloaded". Server is currently experiencing too many requests. Try again and try again later.
503	C	Web browser error for "Service Unavailable". Server is not available due to site is down, moved, too busy, or other issue.
802.11	C	Standard Wireless Networking protocol. IT Directors must decide between 802.11a, 802.11b, and 802.11g.

Suggest New Words http://www.IDII.com/addword.htm

© 2004 Industrial Data & Information Inc.

Glossary of Supply Chain Terminology

Word	Type	Definition of Word
802.11a	C Std	Wireless network standard that operates at 5 GHz with a maximum throughput of 54 Mbps. Range is 80 feet and supplies 12 separate non-overlapping channels. Good for high performance in a small, limited area. Incompatible with existing 802.11b and 802.11g equipment.
802.11b	C Std	Wireless network standard that operates at the crowded 2.4 GHz with a maximum throughput of 11 Mbps. Range is 328 feet and supplies 3 separate non-overlapping channels. Equipment is low cost and available. Compatible with existing 802.11g equipment.
802.11g	C Std	Wireless network standard that operates at the crowded 2.4 GHz with a maximum throughput of 54 Mbps. Range is 328 feet and supplies 3 separate non-overlapping channels. Interoperates with existing 802.11b equipment. Good for larger campuses.
A2A	B C	Application to Application. Connecting software application to another software application. See also EAI, B2B, XML, and EDI.
ADC	C	Automated Data Collection.
Advance Shipping Notification	C	Abbreviated ASN. Details on products being shipped with carton id numbers. May be transmitted by EDI, XML, and/or data on a barcode/RFID tag. See 856 for EDI format.
AES	C Std	Advanced Encryption Standard. See also DES & WEP.
AI	C	Artificial Intelligence. Specialized Computer software to rationally make educated decisions.
AIDC	C	Automatic Identification and Data Control. See AIM.

Suggest New Words http://www.IDII.com/addword.htm

© 2004 Industrial Data & Information Inc.

Glossary of Supply Chain Terminology

Word	Type	Definition of Word
Alphanumeric	C	A character set that contains letters, numbers, and other groups of symbols. (KA)
AP	C	Access Point. A computer hardware device with an antenna to permit wireless users (E.G., RF user) to roam freely among access points. The access point serves as a communications point for wireless clients and provides access to computer databases, electronic messaging, and so forth. Each access point has a connection to a wired LAN.
API	C	Application Programming Interface. A series of programs used to bring data in and out of the database. These programs are developed and maintained by the software application vendor to support interfacing their application to external applications.
Application Package	C	A computer program or set of programs designed for a specific purpose. These can be custom designed, such as for inventory control or forecasting, or "off the shelf" such as spreadsheets or word processors. (KA)
APS	C	Advanced Planning and Scheduling software module. A Supply Chain Planning (SCP) tool to model supplier capacity, profiles, and products.
ASN	C	Advance Shipping Notification. Details on products being shipped with carton id numbers. May be transmitted by EDI, XML, and/or data on a barcode/RFID tag. See 856 for EDI format.
ASP	C	Application Service Provider. Provides application software on their computer systems.
Async	C	Abbreviation for Asynchronous.

Suggest New Words   http://www.IDII.com/addword.htm

© 2004 Industrial Data & Information Inc.

Glossary of Supply Chain Terminology

Word	Type	Definition of Word
Baud	c	A measurement of the signaling speed of a data transmission device; equivalent to the maximum number of signaling elements, or symbols, per second that are generated; may be different from bit/second rat, however, especially at higher speeds, as several bits may be encoded per symbol, or baud with advance encoding techniques such as phase-shift keying.
BIOS	c	Basic Input Output System
Bisync	c	Abbreviation for Bisynchronous.
BizTalk	c	A XML standard for business commerce promoted & supported by Microsoft. Other XML standards include OAGIS by OAG and Rosetta Net.
Bluetooth	c	A wireless protocol for transferring data & information between two devices.
BOB	c	Best of Breed. Specialized software that has advanced functionality and utilizes "best practices".
BPS	c	Bits Per Second. A data transmission rate normally associated with modems and other handheld devices. See Baud Rate.
CAD	c	Computer Aided Design. Computer Aided Drafting.
CAE	c	Computer Aided Engineering (EM)
CASE	c	Computer Aided Software Engineering.
CGI	c	Common Gateway Interface. A program to take information provided from a web page and processes it. A programmer can write a CGI application in a number of different languages such as C, C++, Java, and Perl. Alternatives to CGI are Microsoft's ASP and Macromedia's Cold Fusion (CFM).
Check Digit	c	In bar codes or data processing, a character added to ensure accuracy. (KA)
CICS	c	Customer Information Control System.

Suggest New Words http://www.IDII.com/addword.htm

© 2004 Industrial Data & Information Inc.

Glossary of Supply Chain Terminology

Word	Type	Definition of Word
CIDX	C	Chemical Industry Data Exchange is a data exchange standard based utilizing XML. Used for buying, selling and delivery of chemicals, also known as "Chem eStandards".
Client	C	Any computer connected on a network that requests services from another computer on the network. Normally the client is a PC and is retrieving data from the server (another computer) that has the primary database(s) on it.
COBOL	C	Common Business Oriented Language
COLD	C	Computer Output to Laser Disk. A method to take computer generated reports directly to a document imaging system.
Cookie	C	A small data file created by a Web site while one is browsing on the Internet and stored on your local PC. The browser stores the data in a text file (a human readable file). This data file is then sent back to the web site's server every time the browser requests a page from the web site. The main design of cookies is to identify users and customized web pages for these users.
CPI	C	Characters Per Inch. On a printer, how many characters can print per inch in a printed line. 10 CPI is normal.
CPS	C	Characters Per Second
CPU	C	Central Processing Unit. The CPU is the "brain" of the computer and does all of the processing & calculations. For PC's, Intel and AMD sell chips that are the CPU. The CPU is located on the main circuit board (the motherboard) in a PC.

Suggest New Words   http://www.IDII.com/addword.htm

© 2004 Industrial Data & Information Inc.

Glossary of Supply Chain Terminology

Word	Type	Definition of Word
CRM	C	Customer Relationship Management software module. Manage customer information for personalized service. See also ERM, PRM, and SRM.
CRT	C	Cathode Ray Tube. A computer screen used for data entry. (KA)
CSS	C	Cascading Style Sheets. Using cascading style sheets in web pages for a website, will save time & bring forth consistency for font, font size, font color, and more.
CSV	C	Comma Separated Values. A data file, where fields are separated ("delimited") by commas.
Custom Software	C	Software specially developed for or by the user. (KA)
Data Base	C	Abbreviated DB. Data stored in a from that allows for flexible sortation and report generation. Common data bases include customer lists, routes, carrier selection maps, rate files, and inventory files. (KA)
DB	C	Data Base. E.G., Oracle, Sybase, Informix, SQL Server, Progress.
DBA	C	Data Base Administrator. An IT person that is highly trained to keep the databases operational. Also responsible for backups and optimization of those databases.
DBMS	C	Data Base Management System. A computer application dedicated for the storage of data. The DBMS is organized into databases, tables, columns, rows, and indexes. In older terminology, columns were called fields and rows where called records. AKA DB.
DDE	C	Dynamic Data Exchange.
DES	C Std	Data Encryption Standard. See also AES & WEP.

Suggest New Words http://www.IDII.com/addword.htm

© 2004 Industrial Data & Information Inc.

Glossary of Supply Chain Terminology 41

Word	Type	Definition of Word
Dial-Up	C	A term used to indicate that computers are communicating though modems on standard telephone lines rather than via a dedicated communication link (such as a network or leased telephone lines). (KA)
DLL	C	Dynamic Link Library.
DOS	C	Disk Operating System. One of the first operating systems for PC's was DOS by IBM & Microsoft. DOS is obsolete. Some computer geeks during those early PC years, called DOS another acronym, "Dumb Operating System". See Windows.
DPI	C	Dots Per Inch
DRAM	C	Dynamic Random Access Memory
DSS	C	Decision Support System.
DTD	C	Document Type Definition. RosettaNet has developed an XML standard, using PIPs & DTDs.
Dumb Terminal	C	See Green Screen.
EAI	C	Enterprise Application Integration. E.G., Interfacing data between ERP and WMS. Also, mapping between a WMS and TMS.
ebXML	C	Electronic business eXtensible Mark-up Language. ebXML is an evolutionary framework to accommodate various types of e-transaction technologies ranging from EDI to XML. It is a joint initiative of the UN and OASIS, to advance electronic business by promoting open, collaborative development of interoperability specifications.
EDI	C EDI	Data format specification. See the Electronic Data Interchange section in this glossary.

Suggest New Words http://www.IDII.com/addword.htm

© 2004 Industrial Data & Information Inc.

Glossary of Supply Chain Terminology

Word	Type	Definition of Word
EIP	C	Enterprise Information Portal. Software for an enterprise website(s) offering a user-friendly interface to gain access to enterprise data. EIPs combine search, content management, and DB technologies to deliver information tailored for each user. EIP solutions include BroadVision, Northern Light Technology, and others. EIP can draw from the ERP database, but will also draw information from it's own content management database. Therefore EIP is –not- an ERP system, but a content delivery system to automate common user information requests.
EIS	C	Executive Information System software module. Other terms similar to this are BI, BPM, EPM, and OLAP.
Electronic Data Interchange	C EDI	Computer-to-computer communication between two or more companies that such companies can use to generate bills of lading, purchasing orders, and invoices. It also enables firms to access the information systems of suppliers, customers, and carriers, and to determine the up-to-the-minute status of inventory, orders, and shipments.
EPM	C	Enterprise Performance Management. Software to show, plan, and adjust the performance of the entire enterprise. The EPM must connect to the current enterprise software (ERP, SCM, WMS, TMS) to evaluate details and produce performance indicators.

Suggest New Words   http://www.IDII.com/addword.htm

© 2004 Industrial Data & Information Inc.

Glossary of Supply Chain Terminology 43

Word	Type	Definition of Word
Ethernet	C	Ethernet is a very common LAN access method. The network administrator may configure the Ethernet LAN to contain many types of computer hardware on it, including various computers, printers, access points, hubs, routers, and more. The computers may be running the same or different operating systems and all be connected on the same Ethernet LAN. This permits one PC user to access multiple computers & devices, if his security levels permit such.. See IEEE, 802.11, 802.11a.
EULA	B C	End User License Agreement. A contractual agreement for usage of a software package or operating system.
FDDI	C	Fiber Distributed Data Interface
Firmware	C	Memory in a computer that retains information when the power is off. The memory may be read-only (ROM), programmable read-only (PROM), erasable programmable read-only (EPROM), or electrically erasable read-only (EEPROM). (KA)
Fixed Beam Scanner	C	A barcode scanner that is stationary and reads codes as items move past it. Also Fixed-Position Scanner. (KA)
Fixed Position Scanner	C	See Fixed Beam Scanner. (KA)
FoIP	C	Fax over Internet Protocol. A cost saving method to add fax transmission to broadband networks (intranet and internet). See also VoIP.
FTP	C	File Transfer Protocol. A way to exchange files between computers.

Suggest New Words http://www.IDII.com/addword.htm

© 2004 Industrial Data & Information Inc.

Glossary of Supply Chain Terminology

Word	Type	Definition of Word
GB	C	Gigabyte
Green Screen	C	A character based terminal that does not support a graphical user interface (GUI). The early character based terminals had dark green background with white letters – therefore the term "green screen". Character based terminals were very common in 1960's through the 1980's. Know commonly known as a "dumb terminal".
GUI	C	Graphical User Interface.
Hand-Held Scanner	C	A small, portable scanner that can read the bar code symbol. A wand reader is an example of a hand-held scanner. (KA)
Hard Disk	C	An information storage medium built into computers that can store large amounts of data. Also known as a Disk Drive. (KA)
Hardware	C	A computer term that denotes the machinery that makes up a computer system. It includes video screens, memory units, printers, etc. (KA)
HD	C	High Density. Usually refers to the 1.44 Megabyte 3 ½ " floppy disks that are of high density format.
HTML	C	HyperText Markup Language. A tag language used in building web pages. Web browsers take a web page (ending with .html or .htm), interpret the tags, and display the web page. HTML can be viewed in an editor, whether one uses Notepad, Microsoft Word, or a professional website building tool, such as Macromedia's Dreamweaver.
HTTP	C	HyperText Transfer Protocol. The protocol for moving hypertext files (web pages, etc.) over the Internet.
HW	C	Hardware

Suggest New Words   http://www.IDII.com/addword.htm

© 2004 Industrial Data & Information Inc.

Glossary of Supply Chain Terminology   45

Word	Type	Definition of Word
I18N	C	See Internationalization.
IC	C	Integrated Circuit
Internationalization	C	Internationalization is the process of building software so it is not based on the assumptions of one language or country. The internationalized software product can display and process locale-dependent information such as date, time, address, and number formats properly to he user. See also Localization.
I/O	C	Input / Output.
IP	C	Internet Protocol. See IP Address.
IP Address	C	An unique number for a computer or computer device on a TCP/IP network. Files & data messages are routed based on the destination IP address. Compare this to mail service or parcel delivery, where the letter/shipment, is routed until it reaches its correct destination. A sample IP address is 66.221.212.103 that reaches the computer for IDII.com.
ISDN	C	Integrated Services Digital Network.
ISP	C	Internet Service Provider. A company providing access & bandwidth to the internet. ISP's regularly provide login accounts, e-mail accounts, and host websites.
ISV	C	Independent Software Vendor. A VAR, Value Added Reseller of computer software and/or computer hardware.
IT	C	Information Technology. A generic term for all types of computer software, hardware, network, peoples that work with computers, and the impact of such.
ITL	C Tr	International Trade Logistics. Software for managing international shipment & documentation. Some of the leading ITL software solutions are Vastera, G-Log.

Suggest New Words http://www.IDII.com/addword.htm

© 2004 Industrial Data & Information Inc.

Glossary of Supply Chain Terminology

Word	Type	Definition of Word
ITP	C	Information Technology Providers (EM)
J2ME	C	Java 2 Platform, Micro Edition. A development platform that allows programmers to build applications using the Java programming language and related tools. See also JDBC.
JCL	C	Job Control Language
JDBC	C	Java Database Connectivity. Connecting Java programs with data files of any type, normally through JDBC API's.
L10N	C	See Localization.
LAN	C	Local Area Network. See also Network and WAN.
Laser Scanner	C	Bar-code readers that range in size from hand-held units to larger, fixed-beam scanners. Laser scanners are omnidirectional in that the information contained in the barcode an be read regardless of the orientation of the barcode when scanned. (KA)
LED	C	Light Emitting Diode
Linux	C	A free (or low cost) operating system. It is a derivative of Unix. Linux is available & supported from many sources, including Red Hat Software. Website for Red Hat Software is http://www.redhat.com. Many major vendors are now supporting Linux including IBM, Sun, and others.

Suggest New Words http://www.IDII.com/addword.htm

© 2004 Industrial Data & Information Inc.

Glossary of Supply Chain Terminology

Word	Type	Definition of Word
Localization	C	Localization. Localization is often abbreviated as "I10n "because of the 10 letters between the "I "and the "n ". After a product has been internationalized, it is then localized. . Localization is the process of adapting and translating the product into another language to make it linguistically and culturally appropriate for a particular country. Translation is one of the activities during localization.
LPM	C	Lines Per Minute
MAC	C	Macintosh Computer
MAC	C	Media Access Control. For those using wireless 802.11, it is best to use an encryption to keep wireless messages secured, such as AES, DES, MAC, or WEP.
Machine Readable	C	1) Information that can be read directly from a printed format to an electronic form without human intervention, such as in bar-coding. 2) In computer applications, compiled programs. Human-readable object code Is readable by humans only with the help of programming tools. (KA)
Magnetic Strip	C	A type of identification tag that uses a strip of magnetic material attached to a container or to the merchandise itself. The strip is encoded with information that can be read by a magnetic scanner. (KA)
Master File	C	In data processing, a computer file of data used by various programs. For example, the customer master file may be used by the order-entry and billing routines. (KA)
Metadata	C	Metadata is "data about data". Metadata describes the structure and format that the data is in.

Suggest New Words http://www.IDII.com/addword.htm

© 2004 Industrial Data & Information Inc.

Glossary of Supply Chain Terminology

Word	Type	Definition of Word
MFLOPS	C	Millions of Floating Point Operations Per Second
MHZ	C	Mega-Hertz
MICR	B C	Magnetic Ink Character Recognition. Special ink with magnetic abilities to be read by high peed readers. Commonly used in personal & business checks.
Micro-computer	C	A computer that uses a single-chip microprocessor. Microcomputers and mainframe computers have more power, but microcomputers have evolved so that many are nearly as powerful as mini computers. Personal computers are based on microcomputer architecture. See Personal Computer. (KA)
Middleware	C	Middleware software helps separate application software to talk to each other. I.E., Mercator, IBM's MQ series.
Mincomputer	C	A general-purpose computer smaller than a mainframe, but capable of serving a number of users. (KA)
MIPS	C	Millions of Instructions Per Second
MIS	C Org	Management Information Services, Management Information System. The department or division responsible for computer applications, programmers, developers, quality assurance, documentation of applications, and hot-line support staff. AKA IS and IT.
Modem	C	A computer attachment that allows the transmission of information over telephone lines between computers. A modem converts digital signals to analog for transmission and then converts the signals back to digital at the other end of the phone line. Modem is an acronym for **mo**dulator/**dem**odulator. (KA)

Suggest New Words http://www.IDII.com/addword.htm

© 2004 Industrial Data & Information Inc.

Glossary of Supply Chain Terminology

Word	Type	Definition of Word
Mods	C	Abbreviation for Modifications. Computer people frequently use the term "mods" to indicate a minor or major modification is needed. It can used to discuss a specific modification being designed to do specific business functionality, or it can be used to just indicate that the program would need to be changed in order to perform this.
MP	C	Multi-Processor
MPS	C Mfg	Master Production Scheduling.
MS	C Uom	Millisecond (one one-millionth of a second).
NET, .NET	C	Microsoft's .NET initiative.
Network	C	A network of computers connected together. Other computer equipment may be on the network, such as printers, hubs, and switches. A network is a small defined size and setup & administered by a computer geek called a System Administrator or Network Administrator. A network is usually attached to other networks. Sometimes the entire set of networks is called the "network". E.G., the internet is a large network. A popular network method of connectivity is TCP/IP. See also LAN, WAN.
Non-Read	C W	Failure of scanner to read a barcode or RFID tag.
NT	C	See Windows NT.
N-Tier	C	Software program that is distributed among separate computers over a network. N-Tier implies client/server program model. The most common is 3-tier, where the user interface programs are on the user's computer, the business logic is on a central server, and the data is stored in a database on a third computer.

Suggest New Words http://www.IDII.com/addword.htm

© 2004 Industrial Data & Information Inc.

Glossary of Supply Chain Terminology

Word	Type	Definition of Word
OCR	C	Optical Character Recognition. Converts scanned document into text with expectations of highly accurate recognition rate. Depending upon the quality of the original documents, OCR may give you good results.
ODBC	C	Open DataBase Connectivity.
OE	C	Order Entry. Software where customer's quotes & sales orders are entered and managed. AKA OMS.
OLAP	C	Online Analytical Processing. A term for analyzing data into useful information. Other terms similar to this are BI, BPM, EPM, and EIS.
OLE	C	Object Linking and Embedding.
OLTP	C	On Line Transaction Processing.
OMS	C	Order Management System. Software where customer's quotes & sales order are entered and managed. AKA OE.
ONS	C	Object Naming Service.
OO	C	Object Oriented.
Open Systems	C	A phrase that represents a list of operating systems that are supported by many computer manufacturers. Windows, Linux, and Unix, are considered to be 'open systems'.
Operating System	C	Software that interfaces between the user and the hardware, and between applications and the hardware. The operating system allocates the resources of the computer, such as memory and processing time, to applications. Abbreviated OS. (KA)
Order Entry	C	Abbreviated OE. Software where customer's quotes & sales orders are entered and managed. AKA OMS.
OS	C	Operating System. E.G., Unix, AS/400, Windows 2000.

Suggest New Words http://www.IDII.com/addword.htm

© 2004 Industrial Data & Information Inc.

Glossary of Supply Chain Terminology

Word	Type	Definition of Word
Output Devices	C	Part of a computer system's harware used to communicate information. Examples are video monitors, printers, and modems. (KA)
P2P	C	Peer-to-Peer.
Packaged Software	C	Complete software applications sold to a large market and not intended as custom software. (KA) See also COTS.
PC	C	Personal Computer.
PCB	C	Printed Circuit Board
PDA	C	Personal Digital Assistant or Personal Data Assistant.
PDF	C	Portable Document File. A common computerized document format for keeping text, fonts, and images correctly paginated for presentation. Developed by Adobe.
PDT	C W	Portable data terminals. A rugged hand-held computer with the computing power of a stationary computer. Normally equipped with RF (radio frequency) and an antenna.
Ping	C	Packet Inter-Network Groper. An easy way to test if a computer, hub, router, switch, or network printer is ready to be utilized by users on a network. Ping sends a packet to a designated address and waits for a response.
PIP	C	Partner Interface Process. RosettaNet has developed an XML standard, using PIPs & DTDs. PIPs describe the message mapping, message flow, and process interaction between trading partners. All PIP message schemas are defined using Document Type Definitions (DTDs).

Suggest New Words http://www.IDII.com/addword.htm

© 2004 Industrial Data & Information Inc.

Glossary of Supply Chain Terminology

Word	Type	Definition of Word
PKI	C	Public Key Infrastructure. PKI is a specification using private cryptographic key pairs from a trusted authority. The end result is that PKI users can privately & securely exchange data and money on public networks, such as the internet.
PLC	C	Programmable Logic Controller. Automates and controls either material handling or Manufacturing automation equipment. E.G., diverters controlled by PLC.
PML	C	Physical Markup Language. A data transfer specification based in XML. The Auto ID Center is promoting the PML with ePC and ONS.
PPM	C	Pages Per Minute
PROM	C	Programmable Read Only Memory
Protocol	C	A method to communicate data between software applications on one or more computer systems.
QoS	B C	Quality of Service. Describes fast network speeds with quality of service option
Query	C	In data processing, the process of asking a database program for a list of records that fit certain criteria. (KA)
Radio Interrogation Signal	C	The signal produced by the reader for RFID system that causes the tag to transmit the ID.
RAM	C	Random Access Memory. All computers require memory, which is called RAM. Memory in the computer or needed in the computer is expressed in MB (Megabytes).
REDINET	C	EDI network originally developed by Control Data Corporation; operated by Sterling Commerce's Network Services Group since June 1991.

Suggest New Words   http://www.IDII.com/addword.htm

© 2004 Industrial Data & Information Inc.

Glossary of Supply Chain Terminology

Word	Type	Definition of Word
Redundancy	C	Backup capabilities and duplications of a system designed to reduce or eliminate downtime due to breakdowns. (KA)
RF/DC	C	Radio Frequency Data Communications.
RNIF	C	RosettaNet Implementation Framework. RNIF is a set of specifications utilized to implement message exchange between trading partners. RosettaNet is an XML standard, using PIPs & DTDs.
RO	C	Read Only
Roaming	C	Allows users free travel from one access point (AP) area of coverage to another with no loss in connectivity. Utilized in RF and Bluetooth technologies.
ROM	C	Read Only Memory
RPC	C	Remote Procedure Call.
RPM	C	Revolutions Per Minute
RT	C	Real Time
Scanning Equipment	C	Machines used to read and process transactions by reading encoded information and accounting for the transactions. (KA)
SCEM	C	Supply Chain Event Management software module. Provides "event" based notifications via e-mail, screen pop-ups, fax, pager, and wireless message. Events are also known as alerts. May also include event response logic. See SCPM.
SCES	C W	Supply Chain Execution Systems. Software and processes to enable then execution side of fulfilling customer orders and supporting functions. Includes warehouse operations, transportation, with heavy emphasis on inventory & order entry.

Suggest New Words http://www.IDII.com/addword.htm

© 2004 Industrial Data & Information Inc.

Glossary of Supply Chain Terminology

Word	Type	Definition of Word
SCPM	C	Supply Chain Process Management. SCPM provides alerting, alert response logic, inventory visibility, and KPI. SCPM includes SCEM plus visibility, KPI, and non-compliance event management.
SCSI	C	Small Computer System Interface. Hard disks and other devices have SCSI interfaces. Other types of interfaces are common as well. These interfaces let the computer communicate (read / write) to the devices (I.E., Hard disk).
SEF	C	Standard Exchange Format. Other ways to exchange data include ODBC, EDI, and XML.
SFA	B	Sales Force Automation. Old term now. See CRM.
SLA	B C	Service Level Agreement. A written contract where servicing company agrees to adhere to provide service(s) that are at or above a specified service level.
SMP	C	Symmetrical Multi-Processing
SMS	C	Shipping Manifest System. Computer software that rates freight (parcels, letters, LTL, and TL). Software also produces manifests and shipping labels.
SNA	C	Systems Network Architecture.
Software	C	The coding that instructs a computer how to perform specific functions. (KA)
SQL	C Std	Standard Query Language. A common programming language to query and maintain (add, update, and delete) data. SQL varies slightly from database vendor to vendor.
SRM	C	Supplier Relationship Management software module. See also CRM, ERM, and PRM.

Suggest New Words http://www.IDII.com/addword.htm

© 2004 Industrial Data & Information Inc.

Glossary of Supply Chain Terminology

Word	Type	Definition of Word
System	C	An overall set of computer software, hardware, business processes, and people. I.E., the "ERP system" means the ERP software, plus supporting computer hardware, processes utilized in the business with the ERP software, and people that regularly support or utilize this software.
System 34	C	IBM Computer Series. Came out in the 1970's, replaced by the System 36 & 38.
System 36	C	IBM Computer Series. Came out in the 1970's, replaced by the System 38.
System 38	C	IBM Computer Series. Came out in the 1980's, replaced by the AS/400 and the ISeries.
TCP/IP	C	Transmission Control Protocol / Internet Protocol.
TelNet	C	A terminal emulation program for TCP/IP networks. The Telnet program connects your PC to a server on the network. After logging on with a valid username and password, one may have a user session or an administrator session on that remote host computer.
TP	C	Trading Partner. With EDI or XML, the data is coming from or going to a "trading partner". The trading partner is usually a vendor, a carrier, or a bank.
TP	C	Transaction Processing
TPI	C	Tracks Per Inch
TPS	C	Transactions Per Second

Suggest New Words http://www.IDII.com/addword.htm

© 2004 Industrial Data & Information Inc.

Glossary of Supply Chain Terminology

Word	Type	Definition of Word
UI	C	User Interface. How the user will view and interact with the computer application software. Many user interfaces exist - including old-fashioned "green screen" (character based user interface), GUI, web page, RF handheld screen, RF forklift screen, PDA screen, and voice & 3D vision user interfaces.
Unix	C	A multi-user operating system running on PC's to large mainframes. Most computer manufacturer's sell & support Unix. See also Linux.
URL	C	Uniform Resource Locator. The address of a web site or a web page. For example, http://www.idii.com is an URL and so is ftp.quarterdeck.com
USB	C	Universal Serial Bus. A single USB port can connect a single device or up to 127 computer devices, such as mice, keyboards, modems, and scanners. USB is popular with making new hardware easy to install via Plug-and-Play, hot plugging, and is available on PC's and Macs.
VAN	C	Value Added Networks. Commonly deployed for EDI transmission.
VAR	C	Value Added Resellers. An older term for resellers & integrators of computer software and/or computer hardware. See ISV.
VAS	C	See Value Added Services.
VDT	C	Video Display Terminal. Older terminals running character based applications (not GUI nor Web browser). E.G., Wyse 60, VT220, VT100.

Suggest New Words http://www.IDII.com/addword.htm

© 2004 Industrial Data & Information Inc.

Glossary of Supply Chain Terminology

Word	Type	Definition of Word
Voice Recognition	C	The ability of a computer to receive, reorganize, and understand voice commands as a means of data collection or process control. (KA)
VoIP	C	Voice over Internet Protocol. A cost saving method to add legacy voice transmission to broadband networks (intranet and internet). See also FoIP.
VPN	C Std	Virtual Private Network.
WAN	C	Wide Area Network. See also Network and LAN.
WAP	C	Wireless Application Protocol. Enabling technology to take information from the Internet to be viewed on a cell phone or PDA. See also WML.
WEP	C Std	Wired Equivalent Privacy. An encryption standard. For those using wireless 802.11, it is best to use an encryption to keep wireless messages secured, such as WEP, AES, DES, or MAC.
Windows 9X	C	Refers to either Windows 95 or Windows 98 operating system by Microsoft. Utilized in 1995 through 1999. Was replaced by Windows 2000 and Windows XP.
Windows 2000	C	An operating system for PC's by Microsoft. Built on Windows NT technology. Now replaced by Windows XP.
Windows NT	C	An operating system for PC's by Microsoft. Designed for client-server (multi-user) operations. Now replaced by Windows 2000 then Windows XP.
Windows XP	C	An operating system for PC's by Microsoft. Designed for client-server (multi-user) operations.

Suggest New Words http://www.IDII.com/addword.htm

© 2004 Industrial Data & Information Inc.

Glossary of Supply Chain Terminology

Word	Type	Definition of Word
WML	C	Wireless Markup Language. WML is a markup language that takes the text portions of Web pages and displays them on PDAs and cell phones. See also WAP.
WORM	C	Write Once Read Many
WSDL	C Std	Web Services Definition Language. See http://www.w3.org/TR/wsdl for more details.
WWAN	C	Wireless Wide Area Network.
WWW	C	World Wide Web.
WYSIWYG	C	What You See Is What You Get. Programs that visually display documents and/or images in the same form as when they are printed.
XSLT	C	eXtensible Stylesheet Language Transformations.
Y2K	B C	Year 2000. Older software kept years in a 2-digit year. Year 2000 compliant software has 4 digit year and leap year issues.

Suggest New Words   http://www.IDII.com/addword.htm

© 2004 Industrial Data & Information Inc.

Conveyor

Terminology & Abbreviations Utilized in Conveyors.

Suggest New Words http://www.IDII.com/addword.htm

© 2004 Industrial Data & Information Inc.

Glossary of Supply Chain Terminology

Word	Type	Definition of Word
Accumulating Conveyor	Conv	Any conveyor designed to allow collection (accumulation) of material. May be roller, live roller, belt, or gravity conveyors. (Fs)
Alligator Lacing	Conv	Lacing attached to the belt with a hammer. (Fs)
Axle	Conv	A non-rotating shaft on which wheels or rollers are mounted. (Fs)
Back Pressure	Conv	The amount of force applied to a package to stop the package or collection of packages. (Fs)
Bag Flattener	Conv	A mounting assembly used to hold one conveyor upside down over another conveyor in order to squeeze or flatten the product. (Fs)
Ball Table	Conv	A group of ball transfers over which flat surface objects may be moved in any direction. (Fs)
Ball Transfer	Conv	A device in which a larger ball is mounted and retained on a hemispherical face of small balls. (Fs)
Bare Pulley	Conv	A pulley that does not have the surface of its face covered (or lagged). (Fs)
Bearing	Conv	A machine part in or on which a shaft, axle, pin or other part rotates. (Fs)
Bed	Conv	That part of a conveyor upon which the load rests or slides while being conveyed. (Fs)
Bed Length	Conv	Length of bed sections only required to make up conveyor excluding pulleys, etc., that may be assembled at ends. (Fs)
Bed Width	Conv	Refers to the overall width of the bed section. (Fs)
Belt	Conv	A flexible band placed around two or more pulleys for the purpose of transmitting motion, power or materials from one point to another. (Fs)

Suggest New Words   http://www.IDII.com/addword.htm

© 2004 Industrial Data & Information Inc.

Glossary of Supply Chain Terminology

Word	Type	Definition of Word
Belt Conveyor	Conv	A moving belt designed to carry merchandise. Belt conveyors are used to move materials between facilities, or between floors of a facility. (KA)
Belt Scraper	Conv	A blade or brush caused to bear against the moving conveyor belt for the purpose of removing material sticking to the conveyor belt. (Fs)
Belt Speed	Conv	The length of belt, which passes a fixed point within a given time. It is usually expressed in terms of 'feet per minute'. (Fs)
Between Rail Width	Conv	The distance between the conveyor frame rails on a roller bed, live roller or gravity type conveyor. Abbreviated BR. Also referred to as (BF) Between Frame. (Fs)
BF	Conv	See Between Frame or Between Rail Width.
Booster Conveyor	Conv	Any type of powered conveyor used to regain elevation lost in gravity roller or wheel conveyor lines. (Fs)
BR	Conv	See Between Rail Width.
Brake Motor	Conv	A device usually mounted on a motor shaft between motor and reducer with means to engage automatically when the electric current is cut off or fails. (Fs)
Brake Rollers	Conv	Air or mechanically operated brakes used underneath roller conveyor to slow down or stop packages being conveyed. (Fs)
Butt Coupling	Conv	Angles or plates designed to join conveyor sections together. (Fs)
Capacity	Conv	The number of pieces, volume, or weight of material that can be handled by a conveyor in a unit of time when operating at a given speed. (Fs)

Suggest New Words   http://www.IDII.com/addword.htm

© 2004 Industrial Data & Information Inc.

Glossary of Supply Chain Terminology

Word	Type	Definition of Word
Casters	Conv	Wheels mounted in a fork (either rigid or swivel) used to support and make conveyors portable. (Fs)
Ceiling Hangers	Conv	Lengths of steel rod, attached to the ceiling, from which conveyors may be supported to provide maximum utilization of floor space or when required height exceeds floor support capability. (Fs)
Center Drive	Conv	A drive assembly mounted underneath normally near the center of the conveyor, but may be placed anywhere in the conveyor length. Normally used in reversing or incline application. (Fs)
C Face Drive	Conv	A motor and reducer combination where the two units are flanged and are coupled for connection to each other and have one out-put shaft. (Fs)
Chain	Conv	A series of links pivotally joined together to form a medium for conveying or transmitting motion or power. (Fs)
Chain Conveyor	Conv	Any type of conveyor in which one or more chains act as the conveying element. (Fs)
Chain Drive	Conv	A power transmission device employing a drive chain and sprockets. (Fs)
Chain Guard	Conv	A covering or protection for drive or conveyor chains for safety purposes. (Fs)
Chain Roller Conveyor	Conv	A conveyor in which the tread rollers have attached sprockets which are driven by chain. (Fs)
Chute	Conv	A trough through which objects are lowered by gravity. Can either be a slider bed or roller/wheel bed. (Fs)
Cleat	Conv	An attachment fastened to the conveying surface to act as a pusher, support check or trip, etc. to help propel material, parts or packages along the normal path of conveyor travel. (Fs)

Suggest New Words   http://www.IDII.com/addword.htm

© 2004 Industrial Data & Information Inc.

Glossary of Supply Chain Terminology

Word	Type	Definition of Word
Cleated Belt	Conv	A belt having raised sections spaced uniformly to stabilize flow of material on belts operating on inclines. Cleats may be a part of the belt or fastened on. (Fs)
Clipper Lacing	Conv	Lacing attached to the belt with a clipper-lacing machine. (Fs)
Clutchbrake Drive	Conv	Drive used to disengage motor from reducer and stop conveyor immediately without stopping the motor or cutting the power. (Fs)
Clutch Drive	Conv	Drive used to disengage motor from reducer without stopping the motor or cutting the power. (Fs)
Constant Speed Drive	Conv	A drive with no provisions for variable speed or a drive with the characteristics necessary to maintain a constant speed. (Fs)
Converging	Conv	A section of roller or wheel conveyor where two conveyors meet and merge into one conveyor. (Fs)
Conveying Surface	Conv	Normal working surface of the conveyor. (Fs)
Cross Bracing	Conv	Rods and turnbuckles placed diagonally across roller bed or live roller type conveyors to aid in squaring frames, necessary for tracking purposes. (Fs)
Crossover	Conv	A short section of conveyor placed in a conveyor when drive is switched to opposite side of conveyor. (Fs)
Crowned Pulley	Conv	A pulley that tapers equally from both ends toward the center, the diameter being the greatest at the center. (Fs)
Curve Conveyor	Conv	Any skatewheel, roller or belt conveyor that is produced with a degree of bend so as to convey products away from the straight flow. (Fs)

Suggest New Words   http://www.IDII.com/addword.htm

© 2004 Industrial Data & Information Inc.

Glossary of Supply Chain Terminology

Word	Type	Definition of Word
Decline Conveyor	Conv	A conveyor transporting down a slope. (Fs)
Decree of Incline	Conv	Angle of slope (in degrees) that a conveyor is installed. (Fs)
Differential Curve	Conv	A curved section of roller conveyor having a conveying surface of two or more concentric rows of rollers. Also referred to as a Split Roller design. (Fs)
Discharge End	Conv	Location at which objects are removed form the conveyor. (Fs)
Diverging	Conv	A section of roller or wheel conveying which makes a connection for diverting articles from a main line to a branch. (Fs)
Drive	Conv	An assembly of the necessary structural, mechanical and electrical parts that provide the motive power for a conveyor, usually consisting of motor/reducer, chain, sprockets, guards, mounting base and hardware. (Fs)
Drive Pulley	Conv	A pulley mounted on the drive shaft that transmits power to the belt with which it is in contact. Pulley is normally positive crowned and lagged. (Fs)
Dutchman	Conv	A short section of belt provided with lacing, in a conveyor belt that can be removed when take-up provision has been exceeded. (Fs)
Emergency Pull Cord	Conv	Vinyl coated cord that runs along the side of the conveyor that can be pulled at any time to stop the conveyor. Used with an Emergency Stop Switch. (Fs)
Emergency Stop Switch	Conv	Electrical device used to stop the conveyor in an emergency. Used with an emergency pull cord. (Fs)

Suggest New Words   http://www.IDII.com/addword.htm

© 2004 Industrial Data & Information Inc.

Glossary of Supply Chain Terminology

Word	Type	Definition of Word
Extendable Conveyor	Conv	Roller or wheel conveyor that may be lengthened or shortened within limits to suit operating needs. Standard extended lengths are 20 ft., 30 ft., and 40 ft. (Fs)
EZLogic	Conv	Electronic Zero Pressure Logic - See Hytrol EZLogic Components Manual. (Fs)
Feeder	Conv	A conveyor adapted to control the rate of delivery of packages or objects. (Fs)
Flapper Gate	Conv	A hinged or pivoted plate used for selectively directing material handled. (Fs)
Flat Face Pulley	Conv	A pulley on which the face is a straight cylindrical drum, i.e. uncrowned. (Fs)
Floor Supports	Conv	Supporting members with vertical adjustments for leveling the conveyor. (Fs)
Flow	Conv	The direction of travel of the product on the conveyor. (Fs)
Frame	Conv	The structure that supports the machinery components of a conveyor. (Fs)
Frame Spacer	Conv	Cross members to maintain frame rail spacing. AKA as Bed Spacer. (Fs)
Gate	Conv	A section of conveyor equipped with a hinge mechanism to provide an opening for a walkway, etc., manual or spring-loaded. (Fs)
Gearmotor	Conv	A unit which creates mechanical energy from electrical energy and which transmits mechanical energy through the gearbox at a reduced speed. (Fs)
Gravity Bracket	Conv	Brackets designed to permit gravity conveyors to be attached to the ends of a powered conveyor. (Fs)
Gravity Chute	Conv	A chute or trough used to convey commodities by gravity. (KA)
Gravity Conveyor	Conv	Roller or wheel conveyor over which objects are advanced manually by gravity. (Fs)

Suggest New Words http://www.IDII.com/addword.htm

© 2004 Industrial Data & Information Inc.

Glossary of Supply Chain Terminology

Word	Type	Definition of Word
Guard Rail	Conv	Members paralleling the path of a conveyor and limiting the objects or carriers to movement in a defined path. (Fs)
Hog Rings	Conv	Rings used to hold the shaft in a roller. (Fs)
Horizontal Floor Space	Conv	Floor space required for a conveyor. (Fs)
Hydrostatic Roller Conveyor	Conv	A section conveyor in which the rollers are weighted with liquid to control their rotation and thereby the speed of movement. (KA)
Incline Conveyor Length	Conv	Determined by the elevation change from infeed to discharge versus the degree of incline. (Fs)
Inclined Conveyor	Conv	A conveyor transporting up a slope. (Fs)
Infeed End	Conv	The end of a conveyor nearest the loading point. (Fs)
Intermediate Bed	Conv	A middle section of conveyor not containing the drive or tail assemblies. (Fs)
Interpolate	Conv	To compute intermediate values. (Fs)
Knee Braces	Conv	A structural brace at an angular position to another structural component for the purpose of providing vertical support. (Fs)
Knurl Thumb Adj. Nut	Conv	A nut used on accumulating conveyors to adjust the pressure required to drive the product, may be turned with-out the use of tools. (Fs)
Lacing	Conv	Means used to attach the ends of a belt segment together. (Fs)

Suggest New Words   http://www.IDII.com/addword.htm

© 2004 Industrial Data & Information Inc.

Glossary of Supply Chain Terminology

Word	Type	Definition of Word
Lagged Pulley	Conv	A pulley having the surface of its face crowned with a material to provide for greater friction with the belt. (Fs)
Lagging	Conv	A material applied to the outer surface of pulleys or rollers. (Fs)
Limit Switch	Conv	Electrical device used to sense product location. (Fs)
Live Roller Conveyor	Conv	A series of rollers over which objects are moved by the application of power to all or some of the rollers. The power-transmitting medium is usually belting or chain. (Fs)
Machine Crowned Pulley	Conv	A pulley in which the crown or vertex has been produced by an automatic, usually computer driven, machine. (Fs)
Magnetic Starter	Conv	An electrical device that controls the motor and also provide overload protection to the motor. (Fs)
Manual Start Switch	Conv	A simple one-direction switch used to turn the conveyor on or off. (Fs)
Minimum Pressure Accumulating Conveyor	Conv	A type of conveyor designed to minimize build-up of pressure between adjacent packages or cartons. (I38-ACC - 19O-ACC) (Fs)
Motor	Conv	A machine that transforms electric energy into mechanical energy. Standard motors are dual voltage and operate at 1725 RPM. (Fs)
Negative Crowned Pulley	Conv	A pulley with raised areas set equally in from each end. This crown is used on tail pulleys 24 in. OAW and wider, and aids in belt tracking. (Fs)
Net Lift	Conv	The net vertical distance through which material is moved against gravity by a conveyor. (Fs)

Suggest New Words http://www.IDII.com/addword.htm

© 2004 Industrial Data & Information Inc.

Glossary of Supply Chain Terminology

Word	Type	Definition of Word
Nip Point Guard	Conv	A guard placed to eliminate points or areas on the conveyor where injuries might occur. (Fs)
Noseover	Conv	A section of conveyor with transition rollers placed in conveyor to provide transition from incline to horizontal or horizontal to incline. (Fs)
Nose Roller	Conv	A small roller, used on power belt curve conveyors, to reduce the gap at the transfer points. (Fs)
OAL	Conv	See Overall Length.
OAW	Conv	See Overall Width.
O-Ring	Conv	Polyurethane bands polyurethane used to transmit drive power from roller to roller or spool to roller. (138-SP, 190-SP) (Fs)
Overall Length	Conv	The dimension outside of pulley to outside of pulley including belting or lagging, of any conveyor lengthwise. Abbreviated OAL. (Fs)
Overall Width	Conv	The dimension outside to outside of frame rails. Abbreviated OAW. (Fs)
Overhead Drive	Conv	A drive assembly mounted over a conveyor, which allows clearance for the product. (Fs)
Package Stop	Conv	Any of various devices, either manual or mechanical, used to stop flow on a conveyor. (Fs)
Parts Conveyor	Conv	A conveyor used to catch and transport small parts, stampings, or scrap away from production machinery to hoppers, drums, or other operations. (PC, PCA, PCX, PCH) (Fs)
Photo Cell	Conv	Electrical device used to sense product location. (Fs)
Pivot Plate	Conv	The gusset that attaches the conveyor to the support leg. (Fs)

Suggest New Words http://www.IDII.com/addword.htm

© 2004 Industrial Data & Information Inc.

Glossary of Supply Chain Terminology

Word	Type	Definition of Word
Plastisol Coating	Conv	Poly-vinyl chloride (PVC) covering for roller tubes to prevent product damage or marking. Usually (#70 durometer) green or (#90 durometer) red in color. (Fs)
Plow	Conv	A device positioned across the path of a conveyor at the correct angle to discharge or deflect objects. (Fs)
Poly-Tier Support	Conv	Supporting members capable of supporting more than one level of conveyor at a time. Each tier has vertical adjustment for leveling the conveyor. (Fs)
Pop-Out Roller	Conv	A roller, normally placed on the ends of a belt conveyor used to aid in transfer, and set in a wide groove to allow it to eject if an object comes between it and the belt. (Fs)
Portable Conveyor	Conv	Any type of transportable conveyor, usually having supports that provide mobility. (Fs)
Portable Support	Conv	Supporting members that provide conveyor mobility by use of casters or wheels. (Fs)
Positive Crowned Pulley	Conv	A pulley that tapers equally from both ends toward the center, the diameter being the greatest at the center. The crown aids in belt tracking. (Fs)
Power Belt Curve	Conv	A curve conveyor that utilizes a belt, driven by tapered pulleys. (Fs)
Power Conveyor	Conv	Any type of conveyor that requires power to move its load. (Fs)
Powered Feeder	Conv	A driven length of belt conveyor, normally used to move product horizontally onto an incline conveyor. (Fs)
Pressure Roller	Conv	A roller used for holding the driving belt in contact with the load carrying rollers in a belt driven live roller conveyor. (Fs)

Suggest New Words http://www.IDII.com/addword.htm

© 2004 Industrial Data & Information Inc.

Glossary of Supply Chain Terminology

Word	Type	Definition of Word
Product Footprint	Conv	The surface of the product that comes in contact with the belt rollers, or wheels of the conveyor. (Fs)
Proximity Switch	Conv	An electrical device to control some function of a powered conveyor. (Fs)
Pulley	Conv	A wheel, usually cylindrical, but polygonal in cross section with its center bored for mounting on a shaft. (Fs)
Push Button Station	Conv	An electrical device that operates a magnetic starter. (Fs)
Pusher	Conv	A device, normally air powered, for diverting product from one conveyor line to another line, chute, etc. (Fs)
Rack	Conv	As in a curve, to bend the frame racks down in order that the discharge elevation is lower than the infeed elevation. (Fs)
Receiving End	Conv	The end of a conveyor from which the products move toward you. (Fs)
Return Idler	Conv	A roller that supports the return run of the belt. (Fs)
Reversible	Conv	A conveyor that is designed to move product in either direction. (Fs)
Roller	Conv	A round part free to revolve about its outer surface. The face may be straight, tapered or crowned. Rollers may also serve as the rolling support for the load being conveyed. (Fs)
Roller Bed	Conv	A series of rollers used to support a conveying medium. (Fs)
Roller Centers	Conv	The distance measured along the carrying run of a conveyor from the center of one roller to the center of the next roller. (Fs)
Roller Conveyor	Conv	A series of rollers supported in a frame over which objects are advanced manually, by gravity or by power. (Fs)

Suggest New Words http://www.IDII.com/addword.htm

© 2004 Industrial Data & Information Inc.

Glossary of Supply Chain Terminology

Word	Type	Definition of Word
SC	Conv	See Sortation Conveyor.
Set High	Conv	Vertical spacing that allows the rollers to be mounted above the frame rails. (Fs)
Set Low	Conv	Vertical spacing that allows the roller to be mounted below the top of the frame rails. (Fs)
Shaft	Conv	A bar usually of steel, to support rotating parts or to transmit power. (Fs)
Sheave	Conv	A grooved pulley wheel for carrying a v-belt. (Fs)
Side Channels	Conv	Members that support the rollers on the side of the conveyor. (Fs)
Side Mounted Drive	Conv	A drive assembly mounted to the side of the conveyor, normally used when minimum elevations are required. (Fs)
Side Tables	Conv	Steel tables attached to either side of conveyor bed to provide working surface close to conveyor. (Fs)
Singulation Mode	Conv	Mode where packages are automatically separated while traveling down the conveyor. (Fs)
Skatewheel Conveyor	Conv	A type of wheel conveyor making use of series of skatewheels mounted on common shafts or axles, or mounted on parallel spaced bars on individual axles. (Fs)
Slat Conveyor	Conv	A conveyor that uses steel or wooden slats mounted on roller chain to transport the product. (Fs)
Slave Drive	Conv	A conveyor drive powered from another conveyor instead of having its own prime power source. (Fs)
Slider Bed	Conv	A stationary surface on which the carrying run of a belt conveyor slides. (Fs)
Slug Mode	Conv	Allows all packages to be released simultaneously. (Fs)

Suggest New Words http://www.IDII.com/addword.htm

© 2004 Industrial Data & Information Inc.

Glossary of Supply Chain Terminology

Word	Type	Definition of Word
Snub Idler	Conv	Any rollers used to increase the arc of contact between a belt and drive pulley. (Fs)
Sortation Conveyor	Conv	A conveyor that is able to sort different packages or products to specific take-away lines. Abbreviated SC. (Fs)
Speed Reducer	Conv	A power transmission mechanism designed to provide a speed for the driven equipment less than that of the prime mover. They are generally totally enclosed to retain lubricant and prevent the entry of foreign material. (Fs)
Spool Conveyor	Conv	A conveyor where power to the rollers is accomplished by o-rings driven by spools on a rotating shaft. (138-SP, 190-SP, 25l4-SP) (Fs)
Spur	Conv	A conveyor section to switch unit loads to and from the mainline. (Fs)
Support	Conv	Arrangement of members used to maintain the elevation or alignment of the conveyors. Supports can take the form of hangers, floor supports, or brackets and can be either stationary or portable. (Fs)
Switch	Conv	(1) Any device for connecting two or more contiguous package conveyor lines; (2) An electrical control device. (Fs)
Tail End	Conv	Usually the end of a conveyor nearest loading point. (Fs)
Tail Pulley	Conv	A pulley mounted at the tail end of a conveyor, its purpose is to return the belt. (Fs)
Take-Up	Conv	The assembly of the necessary structural and mechanical parts which provide the means to adjust the length of belt and chain to compensate for stretch, shrinkage or wear and to maintain proper tension. (Fs)

Suggest New Words http://www.IDII.com/addword.htm

© 2004 Industrial Data & Information Inc.

Glossary of Supply Chain Terminology

Word	Type	Definition of Word
Tangent	Conv	Straight portion after a curve conveyor. (Fs)
Tapered Roller	Conv	A conical conveyor roller for use in a curve with end and intermediate diameter proportional to their distance from center of curve. (Fs)
Tapered Roller Curve	Conv	A curved section of roller conveyor having tapered rollers. (Fs)
Throughput	Conv	The quantity or amount of product moved on a conveyor at a given time. (Fs)
Total Load	Conv	Amount of weight distributed over the entire length of a conveyor. (Fs)
Tracking	Conv	Steering the belt to hold or maintain a desired path. (Fs)
Traffic Cop	Conv	A mechanical or electrical mechanism to prevent collision of objects as they merge from two conveyor lines into a single line. (Fs)
Transfer	Conv	A device or series of devices, usually mounted inside a conveyor section, which uses belts, chains, 0-rings, rollers, or skate-wheels, to move products at right angles to adjacent or parallel conveyor lines. (Fs)
Trash Conveyor	Conv	A conveyor, normally a belt conveyor equipped with high side guards, used in transporting empty cardboard boxes and paper trash away from working areas. (TH) (Fs)
Tread Plates	Conv	Diamond top steel filler plates used to till gap between rollers on roller conveyor. (Fs)
Tripod Support	Conv	Three-legged stand for small roller and skatewheel conveyor. Usually easily moved or aligned to maintain elevation of the conveyor. (Fs)

Suggest New Words http://www.IDII.com/addword.htm

© 2004 Industrial Data & Information Inc.

Glossary of Supply Chain Terminology

Word	Type	Definition of Word
Troughed Bed	Conv	A conveyor designed with a deep trough used for carrying broken glass, cans, wood chips, stampings, etc. Also used in recycling operations. (TR, CRB) (Fs)
Troughing Attachments	Conv	Angles used on belt conveyors to cup the edge of the belt. (Fs)
Turnbuckle	Conv	A link with a screw thread at both ends, used for tightening the rod, normally used in cross bracing. (Fs)
Turning Wheel	Conv	Wheel mounted on an adjustable bracket to help insure proper package orientation. (Fs)
Turntable	Conv	A horizontal, rotateable conveyor mechanism used for transferring objects between conveyors that are in angular relation to one another. (900, 1800, 360E) (Fs)
Two-Pulley Hitch	Conv	A special transition section for moving product from horizontal to incline. (TH) (Fs)
Underside Bed Cover	Conv	Sheet metal used to cover the underneath side of a conveyor. (Fs)
Underside Take-Up	Conv	A take-up section located beneath the bed of a belt conveyor. (Fs)
Undertrussing	Conv	Members forming a rigid framework underneath the conveyor, used for supporting the conveyor. (Fs)
Variable Speed	Conv	A drive or power transmission mechanism that includes a speed changing device. Standard mechanical variable speed ratios 6:1 A.C.; electrical variable speed ratio 10:1. (Fs)
V-Belt	Conv	A belt with a trapezoidal cross section for operation in grooved sheaves permitting wedging contact between the belt sides and groove sides. (Fs)

Suggest New Words http://www.IDII.com/addword.htm

© 2004 Industrial Data & Information Inc.

Glossary of Supply Chain Terminology

Word	Type	Definition of Word
Zero Pressure Accumulating Conveyor	Conv	A type of conveyor designed to have zero build-up of pressure between adjacent packages or cartons. (Fs)

Special Note on Conveyor Terminology

Standard **conveyor terms & definitions** are available for purchase from ANSI and CEMA. Reference "ANSI/CEMA Standard 102 "Conveyor Terms and Definitions." See CEMA's website at **http://www.cemanet.org**.

EDI

Established Terminology Utilized By
EDI Specialists

Glossary of Supply Chain Terminology

Word	Type	Definition of Word
104	EDI	Air Shipment Information
110	EDI	Air Freight Details and Invoice
120	EDI	Vehicle Shipping Order
121	EDI	Vehicle Service
125	EDI	Multilevel Railcar Load Details
126	EDI	Vehicle Application Advice
127	EDI	Vehicle Buying Order
128	EDI	Dealer Information
129	EDI	Vehicle Carrier Rate Update
160	EDI	Transportation Automatic Equipment Identification
161	EDI	Train Sheet
163	EDI	Appointment Schedule Information
204	EDI	Motor Carrier Shipment Information
210	EDI	Motor Carrier Freight Details and Invoice
213	EDI	Motor Carrier Shipment Status Inquiry
214	EDI	Transportation Carrier Shipment Status Message
217	EDI	Motor Carrier Loading and Route Guide
218	EDI	Motor Carrier Tariff Information
250	EDI	Purchase Order Shipment Management Document
300	EDI	Reservation (Booking Request) (Ocean)
301	EDI	Confirmation (Ocean)
303	EDI	Booking Cancellation (Ocean)
304	EDI	Shipping Instructions
309	EDI	U.S. Customs Manifest
310	EDI	Freight Receipt and Invoice (Ocean)
311	EDI	Canadian Customs Information
312	EDI	Arrival Notice (Ocean)
313	EDI	Shipment Status Inquiry (Ocean)
315	EDI	Status Details (Ocean)
317	EDI	Delivery/Pickup Order
319	EDI	Terminal Information
322	EDI	Terminal Operations and Intermodal Ramp Activity

Suggest New Words http://www.IDII.com/addword.htm

© 2004 Industrial Data & Information Inc.

Glossary of Supply Chain Terminology

Word	Type	Definition of Word
323	EDI	Vessel Schedule and Itinerary (Ocean)
324	EDI	Vessel Stow Plan (Ocean)
325	EDI	Consolidation of Goods in Container
326	EDI	Consignment Summary List
350	EDI	U.S. Customs Release Information
352	EDI	U.S. Customs Carrier General Order Status
353	EDI	U.S. Customs Events Advisory Details
354	EDI	U.S. Customs Automated Manifest Archive Status
355	EDI	U.S. Customs Manifest Acceptance/Rejection
356	EDI	U.S. Customs Permit to Transfer Request
357	EDI	U.S. Customs In-Bond Information
358	EDI	U.S. Customs Consist Information
361	EDI	Carrier Interchange Agreement (Ocean)
404	EDI	Rail Carrier Shipment Information
410	EDI	Rail Carrier Freight Details And Invoice
412	EDI	A16Trailer/Container Repair Billing
414	EDI	Rail Car-hire Settlements
417	EDI	Rail Carrier Waybill Interchange
418	EDI	Rail Advance Interchange Consist
419	EDI	Advance Car Disposition
420	EDI	Car Handling Information
421	EDI	Estimated Time of Arrival and Car Scheduling
422	EDI	Shipper's Car Order
423	EDI	Rail Industrial Switch List
425	EDI	Rail Waybill Request
426	EDI	Rail Revenue Waybill
429	EDI	Railroad Retirement Activity
431	EDI	Railroad Station Master File
432	EDI	Rail Description
433	EDI	Railroad Reciprocal Switch File
435	EDI	Standard Transportation Commodity Code Master
436	EDI	Locomotive Information
440	EDI	Shipment Weights

Suggest New Words   http://www.IDII.com/addword.htm

© 2004 Industrial Data & Information Inc.

Glossary of Supply Chain Terminology

Word	Type	Definition of Word
451	EDI	Railroad Event Report
452	EDI	Railroad Problem Log Inquiry or Advice
453	EDI	Railroad Service Commitment Advice
455	EDI	Railroad Parameter Trace Registration
456	EDI	Railroad Equipment Inquiry or Advice
460	EDI	Price Distribution or Response Format
463	EDI	Rail Rate Reply
466	EDI	Rate Request
468	EDI	Rate Docket Journal Log
475	EDI	Rail Route File Maintenance
485	EDI	Rate making Action
486	EDI	Rate Docket Expiration
490	EDI	Rate Group Definition
492	EDI	Miscellaneous Rates
494	EDI	Scale Rate Table
503	EDI	Pricing History
504	EDI	Clauses and Provisions
601	EDI	Shipper's Export Declaration
602	EDI	Transportation Services Tender
622	EDI	Intermodal Ramp Activity
715	EDI	Intermodal Group Loading Plan
810	EDI	Invoice.
816	EDI	Organizational Relationships
820	EDI	Payment Order / Remittance Advice.
830	EDI	Planning Schedule With Release Capability
832	EDI	Price/Sales Catalog.
836	EDI	Contract Award.
840	EDI	Request For Quotation.
841	EDI	Specifications / Technical Information.
843	EDI	Response to Request for Quotation.
845	EDI	Price Authorization Acknowledgment/Status
846	EDI	Inventory Inquiry/Advice
847	EDI	Material Claim
850	EDI	Purchase Order.
851	EDI	Asset Schedule
852	EDI	Product Activity Data

Suggest New Words http://www.IDII.com/addword.htm

© 2004 Industrial Data & Information Inc.

Glossary of Supply Chain Terminology

Word	Type	Definition of Word
853	EDI	Routing and Carrier Instruction
854	EDI	Shipment Delivery Discrepancy Information
855	EDI	Purchase Order Acknowledgment.
856	EDI	Ship Notice/Manifest. Advanced Ship Notice that lists goods that were shipped (per carton, pallet, container) and carrier information.
857	EDI	Shipment and Billing Notice
858	EDI	Shipment Information
859	EDI	Freight Invoice
860	EDI	Purchase Order Change Request - Buyer Initiated
861	EDI	Receiving Advice/Acceptance Certificate
862	EDI	Shipping Schedule
864	EDI	Text Message.
865	EDI	Purchase Order Change Acknowledgment/Request - Seller Initiated
866	EDI	Production Sequence
867	EDI	Product Transfer and Resale Report
869	EDI	Order Status Inquiry
870	EDI	Order Status Report
871	EDI	Component Parts Content
875	EDI	Grocery Products Purchase Order
876	EDI	Grocery Products Purchase Order Change
878	EDI	Product Authorization/De-Authorization
879	EDI	Price Change
888	EDI	Item Maintenance
889	EDI	Promotion Announcement
891	EDI	Deduction Research Report
893	EDI	Item Information Request
894	EDI	Delivery/Return Base Record
895	EDI	Delivery/Return Acknowledgment or Adjustment
896	EDI	Product Dimension Maintenance
920	EDI	Loss or Damage Claim - General Commodities
924	EDI	Loss or Damage Claim - Motor Vehicle

Suggest New Words http://www.IDII.com/addword.htm

© 2004 Industrial Data & Information Inc.

Glossary of Supply Chain Terminology

Word	Type	Definition of Word
925	EDI	Claim Tracer
926	EDI	Claim Status Report and Tracer Reply
928	EDI	Automotive Inspection Detail
940	EDI	Warehouse Shipping Order
943	EDI	Warehouse Stock Transfer Shipment Advice
944	EDI	Warehouse Stock Transfer Receipt Advice
945	EDI	Warehouse Shipping Advice
947	EDI	Warehouse Inventory Adjustment Advice
980	EDI	Functional Group Totals
990	EDI	Response to a Load Tender
997	EDI	Functional Acknowledgment. Transaction acknowledgment that trading partner received transaction on this date & time.
998	EDI	Set Cancellation
EDI	C EDI	Data format specification. See Electronic Data Interchange.
EDIFACT	EDI Std	EDI for Administration, Commerce and Transport. EDIFACT is n EDI standard for international usage and was developed by the United Nations.
Electronic Data Interchange	C EDI	Computer-to-computer communication between two or more companies that such companies can use to generate bills of lading, purchasing orders, and invoices. It also enables firms to access the information systems of suppliers, customers, and carriers, and to determine the up-to-the-minute status of inventory, orders, and shipments.
FA	EDI	Functional Acknowledgement. Notice of receipt of an EDI transaction. See 997.
Functional Acknowledgement	EDI	Notice of receipt of an EDI transaction. Abbreviated FA. See 997.

Suggest New Words http://www.IDII.com/addword.htm

© 2004 Industrial Data & Information Inc.

Glossary of Supply Chain Terminology

Word	Type	Definition of Word
GEIS	EDI Org	General Electric Information Services. Provides EDI VAN service. Now known as GE GXS. See http://www.geis.com
GXS	EDI Org	Global eXchange Service. EDI Van service owned by GE. Previously known as GEIS. See http://www.geis.com

Suggest New Words   http://www.IDII.com/addword.htm

© 2004 Industrial Data & Information Inc.

Government

Terms & Words
Found In
Government

Glossary of Supply Chain Terminology

Word	Type	Definition of Word
ADA	Gov Tr	Airline Deregulation Act of 1978. (WA)
BOA	Gov	Basic Order Agreement.
Carmack Amendment	Gov Tr	A federal statute that codifies the common law principle that a carrier is liable for the full value of goods lost, damaged or delayed while in its possession unless the shipper agrees in writing to a lower rate in return for a lower limitation of liability offered by the carrier. (WA)
CFR	Gov	Code of Federal Regulations.
CO	Gov	Contracting Officer.
COTR	Gov	Contracting Officer's Technical Representative.
COTS	B Gov	Commercial Off the Shelf.
Cure Notice	Gov	Notice sent to contractor that contract is in jeopardy and documentation is requested that details how the situation will be remedied.
EO	Gov	Executive Order.
Fair Labor Standards Act	Gov	Abbreviated FLSA. Federal law that governs employment practices such as wages and overtime. (KA)
FAR	Gov	Federal Acquisition Regulation.
FIPS	Gov	Federal Information Processing Standards.
Fitness	Gov	An applicant for certain government licenses must prove that it is a law-abiding citizen with good moral and other character standards. (WA)
FLSA	Gov	See Fair Labor Standards Act. (KA)
FOIA	Gov	Freedom of Information Act.
FSS	Gov	Federal Supply Schedule.

Suggest New Words http://www.IDII.com/addword.htm

© 2004 Industrial Data & Information Inc.

Glossary of Supply Chain Terminology

Word	Type	Definition of Word
GFM	Gov	Government Furnished Material. See also GOCO.
GOCO	Gov	Government Owned / Contractor Operated. See also GFM.
ICCTA	Gov	Interstate Commerce Commission Termination Act of 1995. (WA)
IFB	Gov	Invitation For Bid. See also SIR & SOW.
Matsui Amendment	Gov Tr	49 U.S.C. § 10502 (c)(1) requiring railroads to maintain full value rates on exempt traffic if they offer reduced rates. (WA)
NSN	Gov W	National Stock Number. AKA a SKU, Item, or a Product in private industry.
PP	Gov	Past Performance.
Preemption	Gov	The principle – derived from the Supremacy Clause of the Constitution – that a federal law supersedes any inconsistent state law or regulation. (WA)
Quotas	Gov Tr	Many governments have established quotas of limiting imports by class of goods or country of origin. Sometimes importing countries require issuance of licenses before U.S. companies may ship to them. (Cnf)
RICO	Gov	Racketeer Influenced and Corrupt Organizations Act.
Single Source	Gov	Purposely award a procurement to a single source. (EM)

Suggest New Words http://www.IDII.com/addword.htm

© 2004 Industrial Data & Information Inc.

Glossary of Supply Chain Terminology

Word	Type	Definition of Word
SIR	Gov	Screening Information Request. A request for proposal & information to meet potential acquisition requirements. See also IFB and SOW.
Sole Source	Gov	A non-competitive procurement.
SOW	Gov	Statement of Work. Detailed description of what the Government desires in services and/or products. See IFB & SIR.
SPA	Gov	Simplified Purchasing Agreement.
SSA	Gov	Source Selection Authority.
SSO	Gov	Source Selection Official.
Statutory Notice	Gov Tr	Length of time required by law for carriers to give notice of changes in tariffs, rates, rules and regulations — usually 30 days unless otherwise permitted by authority from ICC or other regulatory body. (Cnf)
TIRRA	Gov	Trucking Industry Regulatory Reform Act.
USC	Gov	United States Code.

Suggest New Words http://www.IDII.com/addword.htm

© 2004 Industrial Data & Information Inc.

Manufacturing

Terms & Words
Found In
Manufacturing

Related Sections: Conveyor, MSDS, and Pallet

Suggest New Words http://www.IDII.com/addword.htm

© 2004 Industrial Data & Information Inc.

Glossary of Supply Chain Terminology

Word	Type	Definition of Word
Bill of Materials	Mfg	A listing of components, parts, and other items needed to manufacture (or assemble) a product.
BOM	Mfg	Bill of Material.
Burden Rate	Mfg	A standard cost added to every production hour to cover overhead expenses. (KA)
CAM	Mfg	Computer Aided Manufacturing.
CERT	Mfg	Certificate of Analysis produced by a laboratory that is testing the product to certify it is within specification.
Certificate of Compliance	Mfg	A certification that suppliers or services meet specified requirements. (KA)
CFPIM	B Mfg	Certified Fellow in Production and Inventory Management. Advanced certification program for Manufacturing and inventory by APICS.
CGMP	Mfg	Current Good Manufacturing Practice. Government regulations. Good Manufacturing Practice regulations (GMPs) are used by pharmaceutical, medical device, and food manufacturers as they produce and test products that people use. See GMP. See http://www.cgmp.com
CGMPs	Mfg	See CGMP.
CIM	Mfg	Computer Integrated Manufacturing.
CPIM	B Mfg	Certified in Production and Inventory Management. Certification program focused on inventory & Manufacturing by APICS.
CRP	Mfg	Capacity Requirements Planning.
CRP	Mfg	Continuous Replenishment Program.
CTO	Mfg W	Configure To Order. See also ETO and MTS.
CTR	Mfg	Cycle Time Reduction

Suggest New Words http://www.IDII.com/addword.htm

© 2004 Industrial Data & Information Inc.

Glossary of Supply Chain Terminology

Word	Type	Definition of Word
Date Code	Mfg W	A label showing the date of production. In the food industry, it becomes an integral part of the lot number. (KA)
Dekitting, Dekit	Mfg W	When a work order has been canceled, the process to return the components back to inventory is called dekitting.
DPM	B Mfg	Defects Per Million.
DPMO	Mfg	Defects per Million Opportunities
DRP	Mfg	Distribution Requirements Planning or Distribution Resource Planning. The distributed version of MRP and MRP II.
ETO	Mfg W	Engineer-To-Order. A product that is designed only when an order requires it. These are usually highly engineered goods and non-commodity items. See also CTO and MTS.
Facility	Mfg W	The physical plant and storage equipment. Permanent storage bins in a warehouse may be considered part of the facility, whereas material handling equipment may not. (KA)
F/G	Mfg W	Finished Goods
Finished Product Inventory	Mfg W	Products available for shipment to customers. AKA Finished Goods. (KA)
GMP	Mfg	Good Manufacturing Practice. Good Manufacturing Practices (GMPs) are US, Canada, and European Government regulations that describe the methods, equipment, facilities, and controls required for producing • Human and veterinary products • Medical devices • Processed food

Suggest New Words   http://www.IDII.com/addword.htm

© 2004 Industrial Data & Information Inc.

Glossary of Supply Chain Terminology

Word	Type	Definition of Word
GMPs	Mfg	Good Manufacturing Practice regulations. See GMP.
HMI	Mfg W	Human-Machine Interface
Hopper	Mfg	A funnel-shaped box that is narrower at the bottom than the top so that it will direct material to a conveyor, feeder, or chute. (KA)
JIT	Mfg	Just In Time. Methodology for reduced inventories, with mindset of zero waste & defects, to arrive to customer at the right time.
Kaizen	Mfg	Japanese term that means continuous improvement. (KA)
Kanban	Mfg	Kanban is a Japanese word for "Just In Time" (JIT) Manufacturing. A Kanban is a signboard or placard used to control the Manufacturing production line.
Lean	Mfg	Short keyword meaning Lean Manufacturing. (ER)
Line Balancing	Mfg	Readjusting product mix and other resources on a production line to achieve the greatest efficiency and consistency. (KA)
Lot-For-Lot	Mfg	Lot sizing technique that matches planned orders with the net requirement for a period. (KA)
Lot Size	Mfg	The amount of a product ordered. It is often equal to EOQ (quantitiy is to be transported in several shipments). Lot size does not include safety stock. (KA)
LTL	Mfg	Lot To Lot.

Suggest New Words http://www.IDII.com/addword.htm

© 2004 Industrial Data & Information Inc.

Glossary of Supply Chain Terminology

Word	Type	Definition of Word
Manufacturing Resource Planning	Mfg	Abbreviated MRP II. MRP II includes MRP plus shop floor, accounting, distribution management, and also activities within Manufacturing and the enterprise. MRP II developed in the 1980's and 1990's. See ERP.
Material Requirements Planning	Mfg	Abbreviated MRP. MRP uses bills of material, inventory data, and MPS to calculate time-phased materials requirements and recommend release or reschedule of orders for materials. Developed in the 1970's. See MRP II.
MES	Mfg	Manufacturing Execution System.
MFG	Mfg	Manufacturing
MPS	C Mfg	Master Production Scheduling.
MPS	Mfg	Master Plan Schedule.
MRP	Mfg	Material Requirements Planning. MRP uses bills of material, inventory data, and MPS to calculate time-phased materials requirements and recommend release or reschedule of orders for materials. Developed in the 1970's. See MRP II.
MRP II	Mfg	Manufacturing Resource Planning. MRP II includes MRP plus shop floor, accounting, distribution management, and also activities within Manufacturing and the enterprise. MRP II developed in the 1980's and 1990's. See ERP.
MS	Mfg	Material Supply
MTBF	Mfg	Mean Time Between Failure. Average running time between failures.
MTF	Mfg	Make To Forecast.
MTO	Mfg W	Make To Order. A product that is manufactured only when an order is confirmed for it. These are usually low volume or highly engineered goods. See also CTO, ETO, and MTS.

Suggest New Words http://www.IDII.com/addword.htm

© 2004 Industrial Data & Information Inc.

Glossary of Supply Chain Terminology

Word	Type	Definition of Word
MTS	Mfg W	Make To Stock. Products are manufactured and placed into warehouse as finished goods. The quantity built is based upon safety stock and re-order point calculations. See also CTO, ETO, and MTO.
MTTR	Mfg	Mean Time To Repair. Average time to perform the repair.
NSP	Mfg	Non-Stock Production system.
OEE	B Mfg	Operating Equipment Effectiveness. Business metric to measure the production line.
PAC	Mfg	Production Activity Control.
PC	Mfg	Production Control
PCS	Mfg	Process Control System (EM)
Phantom Bill	Mfg	A bill of material (BOM) for intermediate manufacturing steps. The results of the phantom bill are used in the next step of manufacturing.
Planning Horizon	Mfg	The amount of time a schedule or forecast extends into the future. (KA)
Production Scheduling	Mfg	The process of scheduling all activities and materials required for manufacturing. The organization of material flows into the production setting in a timely manner. (KA)
Raw Material	Mfg	Goods awaiting conversion into manufactured products or components. (KA)
RCCP	Mfg	Rough-Cut Capacity Planning.
RPI	Mfg	Raw Products Inventory
Scheduled Downtime	Mfg	A planned shutdown of operations to perform maintenance or react to decrease demand. (KA)

Suggest New Words http://www.IDII.com/addword.htm

Glossary of Supply Chain Terminology

Word	Type	Definition of Word
Scrap	Mfg	Excess material, or material that does not meet the required standards, and is unable to be re-worked economically. (KA)
Scrap Factor	Mfg	The percentage of raw materials or components included in gross requirements and expected to be lost in the manufacturing process. (KA)
Semi-Finished Inventory	Mfg W	Materials that are no longer in raw-material form, but which have not completed the production cycle to become finished goods. (KA)
Shop Floor Control	Mfg W	The process of monitoring and controlling production or warehousing activities to ensure that procedures are followed. (KA)
SMM	B Mfg	Small and Medium sized Manufacturers. A grouping of manufacturers by annual revenue size.
SPC	Mfg	Statistical Process Control.
SU	Mfg	Set up. An abbreviation used to denote set-up times or set-up charges. (KA)
SVS	Mfg	Sequence Verification System. Sequencing components and assemblies for the Manufacturing process. A system to control the inventory and sequencing of parts to be utilized.
T&I	Mfg	Testing & Inspection (EM)
VSM	Mfg	Value Stream Mapping.
Waste	Mfg	Residue of material from manufacturing operations that results from mishandling, decay, leakage, shrinkage, etc., and that has no value. Frequently it has a negative value because additional costs and logistical efforts must be incurred for disposal. (KA)
WC	Mfg W	Work Center.

Suggest New Words http://www.IDII.com/addword.htm

© 2004 Industrial Data & Information Inc.

Glossary of Supply Chain Terminology

Word	Type	Definition of Word
WCM	Mfg	World Class Manufacturing. The philosophy of being the best manufacturer of a product. It implies the constant improvement to remain an industry leader.
WIP	Mfg	See Work In Process.
WO	B Mfg	Work Order. An order to build finished or semi-finished goods from a bill of materials.
Work In Process	Mfg	Orders may require value-added services such as price tag printing and application. Once picks for an order a completed they are tracked in WIP staging. See also Value Added Services.
Zero Defects	Mfg W	A long-range objective that strives for defect-free products. (KA)

Suggest New Words http://www.IDII.com/addword.htm

© 2004 Industrial Data & Information Inc.

Glossary of Supply Chain Terminology

Material Safety Data Sheets

Terms & Words
Found In
Material Safety Data Sheets

Suggest New Words http://www.IDII.com/addword.htm

© 2004 Industrial Data & Information Inc.

Glossary of Supply Chain Terminology

Word	Type	Definition of Word
Acute Effect	MSDS	Health effects that usually occur rapidly, as a result of short-term exposure. (HC)
Acute Toxicity	MSDS	Acute effects resulting from a single dose of, or exposure to, a substance. (HC)
Appearance	MSDS	A description of a substance (including color, size, and consistency) at normal room temperature and normal atmospheric conditions. (HC)
Asphyxiant	MSDS	A gas or vapor which can take up space in the air and reduce the concentration of oxygen available for breathing. Examples include acetylene, methane, and carbon dioxide. (HC)
Auto-Ignition Temperature	MSDS	The temperature at which a material will ignite spontaneously or burn. (HC)
Boiling Point	MSDS	Temperature at which a liquid changes to a vapor state at a given pressure (usually sea level pressure = 760 mmHg). (HC)
C or **Ceiling**	MSDS	The maximum allowable human exposure limit for an airborne substance, not to be exceeded even momentarily. Examples: hydrogen chloride, chlorine, nitrogen dioxide, and some isocyanates have ceiling standards.

Suggest New Words http://www.IDII.com/addword.htm

© 2004 Industrial Data & Information Inc.

Glossary of Supply Chain Terminology

Word	Type	Definition of Word
Carcinogen	MSDS	A material that causes cancer. A chemical is considered to be a carcinogen, by OSHA regulation, if: It has been evaluated by the International Agency for Research on Cancer (IARC), and found to be a carcinogen or potential carcinogen; or It is listed as a carcinogen or potential carcinogen in the Annual Report on Carcinogens published by the National Toxicology Program (NTP); or It is regulated by OSHA as a carcinogen; or There is valid scientific evidence in man or animals demonstrating a cancer-causing potential. (HC)
Chronic Health Effects	MSDS	Either adverse health effects resulting from long-term exposure or persistent adverse health effects resulting from short-term exposure. (HC)
Chronic Toxicity	MSDS	Adverse (chronic) effects resulting from repeated doses of or exposures to a substance over a relatively prolonged period of time. Ordinarily used to denote effects in experimental animals. (HC)
Combustible Liquid	MSDS	Any liquid having a flash point at or above 100 °F (37.8 °C), but below 200 °F (93.3 °C), except any mixture having components with flash points of 200 °F (93.3 °C) or higher, the total volume of which make up 99 per cent or more of the total volume of the mixture. (HC)

Suggest New Words http://www.IDII.com/addword.htm

© 2004 Industrial Data & Information Inc.

Glossary of Supply Chain Terminology

Word	Type	Definition of Word
Common Name	MSDS	Any designation or identification such as code name, code number, trade name, brand name, or generic name used to identify a chemical other than by its chemical name. (HC)
Conditions To Avoid	MSDS	Conditions encountered during handling or storage that could cause a substance to become unstable. (HC)
Corrosive Material	MSDS	A liquid or solid that causes visible destruction or irreversible alteration in human skin tissue at the site of contact. (HC)
Decomposition	MSDS	Breakdown of a material or substance (by heat, chemical reaction, electrolysis, decay, or other processes) into simpler compounds. (HC)
Decomposition Products	MSDS	Describes hazardous materials produced during heated operations. (HC)
Density	MSDS	The mass of a substance per unit volume. The density of a substance is usually compared to water, which has a density of 1. Substances which float on water have densities less than 1; substances which sink have densities greater than 1. (HC)
Dermal	MSDS	Used on or applied to the skin. (HC)
Dermal Toxicity	MSDS	Adverse effects resulting from skin exposure to a substance. (HC)
Dry Chemical	MSDS	A powdered, fire-extinguishing agent usually composed of sodium bicarbonate, potassium bicarbonate, etc. (HC)
EHS	MSDS	Environmental Health and Safety Office.

Suggest New Words   http://www.IDII.com/addword.htm

© 2004 Industrial Data & Information Inc.

Glossary of Supply Chain Terminology

Word	Type	Definition of Word
Explosion Limits	MSDS	The range of concentration of a flammable gas or vapor (% by volume in air) in which explosion can occur upon ignition in a confined area. The boundary-line mixtures of vapor or gas with air, which, if ignited, will just propagate the flame. (HC)
Explosive	MSDS	A chemical that causes a sudden, almost instantaneous release of pressure, gas, and heat when subjected to sudden shock, pressure, or high temperature. (HC)
Extinguishing Media	MSDS	Specifies the fire-fighting agents that should be used to extinguish fires. (HC)
Flammable	MSDS	A chemical that includes one of the following categories: Liquid, flammable--Any liquid having a flash point below 100 °F (37.8 °C), except any mixture having components with flash points of 100 °F (37.8 °C) or higher, the total of which make up 99 percent or more of the total mixture volume. Solid, flammable--A solid, other than an explosive, that can cause fire through friction, absorption of mixture, spontaneous chemical change, or retained heat from manufacturing or processing, or that can be readily ignited and, when ignited, will continue to burn or be consumed after removal from the source of Ignition. (HC)

Suggest New Words http://www.IDII.com/addword.htm

© 2004 Industrial Data & Information Inc.

Glossary of Supply Chain Terminology

Word	Type	Definition of Word
Flash Point	MSDS	The temperature at which a liquid will give off enough flammable vapor to ignite. The lower the flash point, the more dangerous the product. A "flammable liquid" is a solution with a flash point below 100 °F (37.8 °C). Flash point values are most important when dealing with hydrocarbon solvents. The flash point of a material may vary depending on the method used, so the test method is indicated when the flash point is given. (HC)
Foreseeable Emergency	MSDS	Any potential occurrence such as, but not limited to, equipment failure, rupture of containers, or failure of control equipment, which could result in an uncontrolled release of hazardous chemical into the testing environment. (HC)
Hazardous Material	MSDS	In a broad sense, any substance or mixture of substances having properties capable of producing adverse effects on the health or safety of a human being. (HC)
Hazard Ratings	MSDS	Material ratings of one to four which indicate the severity of hazard with respect to health, flammability, and reactivity. (HC)
Hazard Warnings	MSDS	Any words, pictures, symbols, or combination thereof appearing on a label or other appropriate form of warning which conveys the hazards of the chemical(s) in the container(s). (HC)

Suggest New Words   http://www.IDII.com/addword.htm

© 2004 Industrial Data & Information Inc.

Glossary of Supply Chain Terminology

Word	Type	Definition of Word
Health Hazard	MSDS	A chemical for which there is statistically significant evidence, based on at least one study conducted in accordance with established scientific principles, that acute or chronic health effects may occur in exposed employees. The term "health hazard" includes chemicals which are carcinogens, toxic or highly toxic agents, reproductive toxins, irritants, corrosives, sensitizers, hepatoxins, nephrotoxins, neurotoxins, agents that can act on the hematopoietic system, and agents which damage the lungs, skin, eyes, or mucous membranes. (HC)
Incompatible	MSDS	Materials that could cause dangerous reactions from direct contact with one another. These types of chemicals should never be stored together. (HC)
Ingestion	MSDS	The taking in of a substance through the mouth. (HC)
Inhalation	MSDS	The breathing in of a substance in the form of a gas, vapor, fume, mist, or dust. (HC)
Irritant	MSDS	A substance which, by contact in sufficient concentration for a sufficient period of time, will cause an inflammatory response or reaction of the eye, skin, or respiratory system. The contact may be a single exposure or multiple exposure. (HC)
LD 50	MSDS	See Lethal Dose 50.

Suggest New Words   http://www.IDII.com/addword.htm

© 2004 Industrial Data & Information Inc.

Glossary of Supply Chain Terminology

Word	Type	Definition of Word
LEL, LFL	MSDS	Lower explosive limit, or lower flammable limit, of a vapor or gas; the lowest concentration (lowest percentage of the substance in air) that will produce a flash of fire when an ignition source (heat, arc, or flame) is present. At concentrations lower than the LEL, the mixture is too "lean" to burn. See UEL.. (HC)
Lethal Dose 50	MSDS	A single dose of a material expected to kill 50 percent of a group of test animals. The dose is expressed as the amount per unit of body weight, the most common expression being milligrams of material per kilogram of body weight (mg/kg of body weight). Usually refers to oral or skin exposure. (HC)
Material Safety Data Sheet	W MSDS	A Material Safety Data Sheet (MSDS) is prepared by a chemical manufacturer, and summarizes available information on the health, safety, fire, and environmental hazards of a chemical product. It also gives advice on the safe use, storage, transportation, and disposal of that product. Other useful information such as physical properties, government regulations affecting the product, and emergency telephone numbers are provided in the MSDS as well. There is a detailed description of how to read an MSDS and a useful glossary of MSDS terms in Hach's Website at http://www.hach.com. (HC)
Melting Point	MSDS	The temperature at which a solid substance changes to a liquid state. For mixtures, the melting range may be given. (HC)

Suggest New Words http://www.IDII.com/addword.htm

© 2004 Industrial Data & Information Inc.

Glossary of Supply Chain Terminology

Word	Type	Definition of Word
Mixture	MSDS	Any combination of two or more chemicals if the combination is not, in whole or in part, the result of a chemical reaction. (HC)
MSDS	W MSDS	See Material Safety Data Sheet.
Mutagen	MSDS	Those chemicals or physical effects that can alter genetic material in an organism and result in physical or functional changes in all subsequent generations. (HC)
NFPA	MSDS Org	National Fire Protection Association is an international membership organization which promotes/ improves fire protection and prevention and establishes safeguards against loss of life and property by fire. Best known on the industrial scene for the National Fire Codes (16 volumes of codes, standards, recommended practices and manuals developed and periodically updated by NFPA technical committees). Among these is NFPA 704M, the code for showing hazards of materials as they might be encountered under fire or related emergency conditions, using the familiar diamond-shaped labels or placards with appropriate numbers and symbols. (HC)
NTP	MSDS	National Toxicology Program. The NTP publishes an Annual Report on Carcinogens which identifies substances that have been studied and found to be carcinogens in animal or human evaluations. (HC)
Oral Toxicity	MSDS	Adverse effects resulting from taking a substance into the body via the mouth. Ordinarily used to denote effects in experimental animals. (HC)

Suggest New Words http://www.IDII.com/addword.htm

© 2004 Industrial Data & Information Inc.

Glossary of Supply Chain Terminology

Word	Type	Definition of Word
OSHA	MSDS Org	Occupational Safety and Health Administration. See http://www.osha.gov See also DOL and NIOSH, as OSHA collaborates with other government agencies for worker safety guidelines & regulations.
Oxidizer	MSDS	A chemical other than a blasting agent or explosive that initiates or promotes combustion in other materials, thereby causing fire either of itself or through the release of oxygen or other gases. (HC)
PEL's	MSDS	See Permissible Exposure Limits.
Permissible Exposure Limits	MSDS	Permissible Exposure Limits (PEL's) are OSHA's legal exposure limits. (HC)
PH	MSDS	A symbol relating the hydrogen ion (H+) concentration of that of a given standard solution. A pH of 7 is neutral. Numbers from 7 to 14 indicate greater alkalinity. Numbers from 7 to 0 indicate greater acidity. (HC)
Physical Hazard	MSDS	A chemical for which there is scientifically valid evidence that it is a combustible liquid, a compressed gas, explosive, flammable, an organic peroxide, an oxidizer, pyrophoric, unstable (reactive) or water-reactive. (HC)

Suggest New Words http://www.IDII.com/addword.htm

© 2004 Industrial Data & Information Inc.

Glossary of Supply Chain Terminology

Word	Type	Definition of Word
Polymerization	MSDS	A chemical reaction in which one or more small molecules combine to form larger molecules at a rate that releases large amounts of energy. If hazardous polymerization can occur with a given material, the MSDS will usually list conditions which could start the reaction. In most cases the material contains a polymerization inhibitor, which if used up, is no longer capable of preventing a reaction. (HC)
PPB	MSDS Uom	Parts Per Billion. Parts of vapor or gas per billion parts of contaminated air by volume. (HC)
PPM	MSDS Uom	Parts Per Million. Parts of vapor or gas per million parts of contaminated air by volume. (HC)
RCRA	MSDS	Resource Conservation and Recovery Act, administered by the EPA.
Reactivity	MSDS	A description of the tendency of a substance to undergo chemical reaction with the release of energy. Undesirable effects such as pressure build-up, temperature increase, and formation of noxious, toxic or corrosive byproducts may occur because of the reactivity of a substance by heating, burning, direct contact with other materials, or other conditions in use or in storage. (HC)

Suggest New Words http://www.IDII.com/addword.htm

© 2004 Industrial Data & Information Inc.

Glossary of Supply Chain Terminology

Word	Type	Definition of Word
SARA Title III	Gov MSDS	Title III of the Superfund Amendments and Reauthorization Act of 1986, also known as the Emergency Planning and Community Right-To-Know Act. It requires extensive submission of information about hazardous chemicals to EPA, states, and local communities, and establishes a national program of emergency planning. Administered by EPA. (HC)
Sensitizer	MSDS	A substance which on first exposure causes little or no reaction, but which on repeated exposure may cause a marked response not necessarily limited to the contact site. Skin sensitization is the most common form of sensitization in the industrial setting, although respiratory sensitization to a few chemicals is also known to occur. (HC)
Shipping Information	MSDS Tr	The appropriate name(s), hazard class(es), and identification number(s) as determined by the US DOT, International Regulations, and the International Civil Aviation Organization. (HC)
Specific Gravity	MSDS	The weight of a material compared to the weight of an equal volume of water is an expression of the density (or heaviness) of a material. Insoluble materials with specific gravity of less than 1.0 will float in or on water. Insoluble materials with specific gravity greater than 1.0 will sink in water. Most (but not all) flammable liquids have specific gravity less than 1.0 and, if not soluble, will float on water - an important consideration for fire suppression. (HC)
Teratogen	MSDS	Any substance that causes growth abnormalities in embryos, genetic modifications in cells, etc. (HC)

Suggest New Words   http://www.IDII.com/addword.htm

© 2004 Industrial Data & Information Inc.

Glossary of Supply Chain Terminology

Word	Type	Definition of Word
Threshold Limit Values	MSDS	Expresses the airborne concentration of a material to which nearly all persons can be exposed, day after day, without adverse effects. TLV's are expressed three ways: (HC)
Time Weighted Average Exposure	MSDS	The airborne concentration of a material to which a person is exposed, averaged over the total exposure time, generally the total workday (8 to 12 hours). Abbreviated as TWA. (HC)
TLV	MSDS	See Threshold Limit Values.
TLV-C	MSDS	The ceiling exposure limit is the concentration that should never be exceeded, even instantaneously. (HC)
TLV-STEL	MSDS	The short-term exposure limit or maximum concentration for a continuous 15-minute exposure period (maximum of four such periods per day, with at least 60 minutes between exposure periods) and provided the TLV-TWA is not exceeded. (HC)
TLV-TWA	MSDS	The allowable Time Weighted Average concentration for a normal 8-hour workday (40-hour work week). (HC)

Suggest New Words http://www.IDII.com/addword.htm

© 2004 Industrial Data & Information Inc.

Word	Type	Definition of Word
Toxic	MSDS	TOXIC: Refers to a chemical falling within any of the following toxic categories: A chemical that has a median lethal dose (LD50) of more than 50 milligrams per kilogram, but not more than 500 milligrams per kilogram of body weight when administered orally to albino rats weighing between 200 and 300 milligrams each. A chemical that has a median lethal dose (LD50) of more than 200 milligrams per kilogram, but not more than 1000 milligrams per kilogram of body weight when administered by continuous contact for 24 hours (or less if death occurs within 24 hours) with the bare skin of albino rabbits weighing between 2 and 3 kilograms each. A chemical that has a median lethal concentration (LC50) in air of more than 200 parts per million, but not more than 2000 parts per million by volume of gas or vapor, or more than two milligrams per liter, but not more than 20 milligrams per liter of mist, fume or dust, when administered by continuous inhalation for one hour (or less if death occurs within one hour) to albino rats weighing between 200 and 300 grams each. (HC)
Toxicity	MSDS	The sum of adverse effects resulting from exposure to a material, generally by the mouth, skin, or respiratory tract. (HC)
TWA	MSDS	See Time Weight Average exposure.
UEL	MSDS	See Upper Explosive Limit.
UFL	MSDS	See Upper Flammable Limit.

Suggest New Words http://www.IDII.com/addword.htm

© 2004 Industrial Data & Information Inc.

Glossary of Supply Chain Terminology

Word	Type	Definition of Word
Unstable	MSDS	Tending toward decomposition or another state, or as produced or transported, will vigorously polymerize, decompose, condense, or become self-reactive under condition of shocks, pressure, or temperature. (HC)
Upper Explosive Limit, Upper Flammable Limit	MSDS	Upper explosive limit (UEL) or upper flammable limit (UFL) of a vapor or gas; the highest concentration (highest percentage of the substance in air) that will produce a flash of fire when an ignition source (heat, arc, or flame) is present. At higher concentrations, the mixture is too "rich" to burn. See LEL. (HC)
Vapor Density	MSDS	The density of a material's vapor compared to the density of the air. If a vapor density is greater than one, it is more dense than air and will drop to the floor or the lowest point available. If the density is less than one, it is lighter than air and will float upwards like helium. (HC)
VOC	MSDS	Volatile Organic Content.
Water Reactive	MSDS	A chemical that reacts with water to release a gas that is either flammable or presents a health hazard. (HC)

Suggest New Words http://www.IDII.com/addword.htm

© 2004 Industrial Data & Information Inc.

Organization

Non-Profit
And For Profit
Organizations

Glossary of Supply Chain Terminology

Word	Type	Definition of Word
3PF	Org	Third Party Fulfillment provider.
3PL	Org	Third Party Logistics provider. A separate company that manages the inventory and logistics of shipments. 3PLs provide transportation management, freight forwarding, customs brokerage, warehousing, kitting, light manufacturing and distribution services. A 3PL can be either asset or non-asset based. Many operations are outsourced to 3PLs.
4PL	Org	Fourth Party Logistics provider. A master contractor who manages an entire outsourced logistics network for a company. 4PLs traits include multi-modal services, advanced technology, international reach, and able to manage complex requirements. AKA as a LLP.
AAIA	Org	Automotive Aftermarket Industry Association (AAIA). See http://www.aftermarket.org/
AAPA	Org	American Association of Port Authorities. See http://www.aapa-ports.org/
AAR	Org	American Association of Railroads. See http://www.aar.org/
AFIF	Org	Association of Floral Importers of Florida. See http://www.siteblazer.net/afif
AHTD	Org	Association for High Technology Distribution. See http://www.ahtd.org/
AIAG	Org	Automotive Industry Action Group. See http://www.aiag.org

Suggest New Words http://www.IDII.com/addword.htm

© 2004 Industrial Data & Information Inc.

Glossary of Supply Chain Terminology

Word	Type	Definition of Word
AIM	Org	Automatic Identification Manufacturers Association. Industry association of the automatic identification and data capture (AIDC) industry. Their mission is to educate end users on AIDC solutions using Bar Codes, RFID, RF Data Communications, and much more. See http://www.aimusa.org
AMR	Org	AMR Research. A research organization for business applications & technology research. Website for AMR Research is http://www.amrresearch.com.
ANLA	Org	American Nursery & Landscape Association. See http://www.anla.org/
APDA	Org	American Parts Distributors Association, Inc. See http://www.apda.com
APEC	Org	Asian-Pacific Economic Caucus. See http://www.apec.org/
APICS	Org	American Production & Inventory Control Society. Educational, non-profit association focused on Manufacturing, inventory, distribution, purchasing, logistics, and supply chain. Has CPIM and CIRM certification programs based upon APIC's body of knowledge. See http://www.apics.org
APRA	Org	Automotive Parts Rebuilders Association. See http://www.apra.org/
ARC	Org	IT research group with focus on Manufacturing, warehousing, and future growth trends. See http://www.arcweb.com
ARPA	Org	Advanced Research Projects Agency. See http://www.darpa.mil/
ASA	Org	American Supply Association. See http://www.asa.net/
ASCD	Org	Association of Service & Computer Dealers International. See http://www.ascdi.com/

Suggest New Words http://www.IDII.com/addword.htm

© 2004 Industrial Data & Information Inc.

Glossary of Supply Chain Terminology

Word	Type	Definition of Word
ASME	Org	American Society of Mechanical Engineers. See http://www.asme.org/
ASPE	Org	American Society of Professional Estimators. See http://www.aspenational.com/
ASQ	Org	American Society for Quality. See http://www.asq.org/
ASRS/AGVS	Org	ASRS/AGVS user's association. See http://www.asrs.org/
ASTL	Org	American Society of Transportation & Logistics. Promotes professionalism and continuing education in the field of transportation and logistics. See http://www.astl.org.
ASTM	Org	American Society for Testing Materials. Develops standards such as DSRC (dedicated short-range communications). See also ITS. See http://www.astm.org/
ATA	Org	Air Transportation Association of America. See http://www.air-transport.org/
ATA	Org	American Trucking Association. See http://www.trucking.org/
ATMI	Org	American Textile Manufacturers Institute. See http://www.atmi.org/
ATRI	Org Tr	American Transportation Research Institute, formerly the ATA Foundation. See http://www.atri-online.org/
AVDA	Org	American Veterinarian Distributors Association. See http://www.avda.net/
AWA	Org	American Warehousemen's Association, which has been renamed to the IWLA.
AWFS	Org	Association of Woodworking and Furnishings Suppliers. See http://www.awfs.org/
AWMA	Org	American Wholesale Marketers Association. See http://www.awmanet.org/

Suggest New Words http://www.IDII.com/addword.htm

© 2004 Industrial Data & Information Inc.

Glossary of Supply Chain Terminology

Word	Type	Definition of Word
BBSI	Org	Beauty & Barber Supply Institute. See http://www.bbsi.org/
CALM	Org	Canadian Association of Logistics Management. Non-profit organization of business professionals interested in improving their logistics management skills. CALM has annual conferences, education, training, and periodical. See http://www.calm.org
CBP	Org	U.S. Bureau of Customs and Border Protection. Cargo manifest rules require shippers to notify the CBP via electronic means before bringing cargo into or send cargo out of the U.S. CBP continues to implement Container Security Initiative (CSI) at major ports around the world. See http://www.cbp.gov
CCRA	Org Tr	Canada Customs and Revenue Agency. See http://www.ccra-adrc.gc.ca/
CEMA	Org	Conveyor Equipment Manufacturers Association. See http://www.cemanet.org for information. CEMA has a number of educational & technical publications on conveyors, terminology, and standards. CEMA & ANSI produce Standard Publication 102 called "Conveyor Terms & Definitions".
CHEP	Org Pa	CHEP founded in 1946 and is a multi-national company known primary for CHEP pallets that it rents. The abbreviation CHEP stands for "Commonwealth Handling Equipment Pool". CHEP is equipment pooling system provider. See http://www.chep.com.

Suggest New Words http://www.IDII.com/addword.htm

© 2004 Industrial Data & Information Inc.

Glossary of Supply Chain Terminology

Word	Type	Definition of Word
CHSCN	Org	Canadian Healthcare Supply Chain Network. "Promoting safe and quality healthcare through the implementation of optimal supply chain management practices and systems" is the vision of the CHSCN. See also EHCR and ECR. See http://www.loginstitute.ca/health_1.html
CIA	Org	Central Intelligence Agency. See http://www.cia.gov
CIFFA	Org	Canadian International Freight Forwarders Association. See http://www.ciffa.com
CITT	Org	Canadian Institute of Traffic & Transportation. See http://www.citt.ca/
CLM	Org	Council of Logistics Management. Educational, non-profit association focused on distribution, logistics, and supply chain. CLM is one of the largest organizations. See http://www.clm1.org.
CTDA	Org	Ceramic Tile Distributors Association. See http://www.ctdahome.org/
CWA	Org	Central Wholesalers Association. See http://www.asaonline.org/
CWDA	Org	Canadian Wholesale Drug Association.
DARPA	Org	Defense Advance Research Project Administration. See http://www.darpa.mil/
D&B	Org	Dun & Bradstreet. US based organization providing financial statistics and credit ratings of businesses. See http://www.db.com.
Department of Transportation	Org	The federal agency that regulates the highway transportation of freight, including commodities designated as hazardous material. See www.dot.gov/
DISA	Org	Data Interchange Standards Association. See http://www.disa.org/
DOC or DoC	Org	U.S. Department of Commerce. See http://www.doc.gov/

Suggest New Words http://www.IDII.com/addword.htm

© 2004 Industrial Data & Information Inc.

Glossary of Supply Chain Terminology

Word	Type	Definition of Word
DOD or DoD	Org	U.S. Department of Defense. See http://www.defenselink.mil/
DOE	Org	U.S. Department of Energy. See http://www.doe.gov and http://www.energy.gov.
DOT	Org	U.S. Department of Transportation. See Department of Transportation. See www.dot.gov/
DOL	Org	U.S. Department of Labor. See http://www.dol.gov.
EC	Org	European Commission or European Community. See http://www.europa.eu.int/
ECA	Org	Express Carriers Association. See http://www.expresscarriers.com/
ECCC	Org	Electronic Commerce Council of Canada. Associated with the UCC Council on promoting standards for commerce. See http://www.eccc.org
EFTA	Org	European Free Trade Association. See http://www.efta.int/
ELA	Org	European Logistics Association. A federation of 36 national organizations, covering almost every country in western Europe. Concerned with logistics within Europe and serves industry and trade. See http://www.elalog.org
ESCF	Org	European Supply Chain Forum. See http://www.tm.tue.nl/efgscm/
FAA	Org	Federal Aviation Agency. US government agency responsible for air traffic and air safety. See http://www.faa.gov/
FCC	Org	Federal Communications Commission. See http://www.fcc.gov/
FDA	Org	U.S. Food & Drug Administration. See http://www.fda.org

Suggest New Words   http://www.IDII.com/addword.htm

© 2004 Industrial Data & Information Inc.

Glossary of Supply Chain Terminology

Word	Type	Definition of Word
FDI	Org	Food Distributors International. See http://www.fdi.org/
FEDA	Org	Foodservice Equipment Distributors Association. See http://www.feda.com/
FEWA	Org	Farmers Equipment Wholesalers Association. See http://www.fewa.org/
FISA	Org	Food Industry Suppliers Association. See http://www.iafis.org/
FMCSA	Org Tr	Federal Motor Carrier Safety Administration. US government agency responsible for overseeing motor carrier safety & guidelines. See http://www.fmcsa.dot.gov/
GAO	Org	General Accounting Office. See http://www.gao.gov/
GCI	Org	Global Commerce Initiative. Started October 1999. A voluntary global user group to facilitate best practice process recommendations, especially in e-commerce and supply chain. See http://www.globalcommerceinitiative.org
GEIS	Org EDI	General Electric Information Services. Provides EDI VAN service. Now known as GE GXS. See http://www.geis.com
GMA	Org	Grocery Manufacturers of America. The GMA advances the interests of the food, beverage, and consumer products industry on key issues that affect the brand manufacturors. See http://www.gmabrands.org
GMDC	Org	General Merchandise Distributors Corporation. See http://www.gmdc.com/
GNDA	Org	Greater North Dakota Association / Wholesalers Division. See http://www.gnda.com/

Suggest New Words http://www.IDII.com/addword.htm

© 2004 Industrial Data & Information Inc.

Glossary of Supply Chain Terminology

Word	Type	Definition of Word
Grocery Pallet Council	Org	A group that provided a standard set of regulations for pallet size and specifications. (KA)
GSA	Org	General Services Administration. See http://www.gsa.gov/
GXS	Org EDI	Global eXchange Service. EDI Van service owned by GE. Previously known as GEIS. See http://www.geis.com
HDMA	Org	Healthcare Distribution Management Association. See http://www.healthcaredistribution.org/
HIA	Org	Hobby Industry of America. See http://www.hobby.org/
HIDA	Org	Health Industry Distributors Association. See http://www.hida.org/
HIDC	Org	Holland International Distribution Council. See http://www.hidc.nl/
I2	Org	Software company providing TMS solutions worldwide. Manugistics and i2 Technologies are considered leaders in the TMS software industry. Leaders in PLANNING systems. See http://www.i2.com.
IAPD	Org	International Association of Plastics Distributors. See http://www.iapd.org/
IARW	Org	International Association of Refrigerated Warehouses. See http://www.iarw.org/
IATA	Org	International Air Transport Association. See http://www..iata.gov/

Suggest New Words http://www.IDII.com/addword.htm

© 2004 Industrial Data & Information Inc.

Glossary of Supply Chain Terminology

Word	Type	Definition of Word
ICC	Org	Interstate Commerce Commission. An independent regulatory agency that implemented federal economic regulations controlling railroads, motor carriers, pipelines, domestic water carriers, domestic surface freight forwarders, and brokers. The I.C.C. legislated out of existence in 1996 with the passage of the Interstate Commerce Commission Termination Act. (WA)
ICPA	Org Tr	International Compliance Professionals Association (ICPA) is an organization for trade compliance professionals. See http://www.int-comp.org/
IDA	Org	Independent Distributors Association. See http://www.idaparts.org/
IEEE	Org	Institute of Electrical and Electronic Engineers. An organization for setting standards for computers and communications. See 802.11, 802.11A. See http://www.ieee.org.
IIPMM	Org	Irish Institute of Purchasing Materials Management. See http://www.iipmm.ie/
IMDA	Org	Independent Medical Distributors Association. See http://imda.org/
Intermodal	Org	Intermodal Association of North America. A transportation method that combines several MODES of transportation in combination to ship something to its destination. I.e. TL, Ocean, TL. See http://www.intermodal.org/
IOLT	Org	Institute of Logistics and Transport. See http://www.iolt.org.uk/
IS	Org	Information Systems. See also IT, MIS.
ISD	Org	Independent Sealing Distributors. See http://www.isd.org/

Suggest New Words http://www.IDII.com/addword.htm

© 2004 Industrial Data & Information Inc.

Glossary of Supply Chain Terminology

Word	Type	Definition of Word
ISO	Org	International Standards Organization. A worldwide federation of national standards bodies from more than 140 countries. ISO was established in 1947. See http://www.iso.ch
ISSA	Org	International Sanitary Supply Association. See http://www.issa.com/
ISTA	Org	International Safe Transit Association. See http://www.ista.org/
ITA	Org	Industrial Truck Association. See http://www.indtrk.org/
IWFA	Org	International Wholesale Furniture Association. See http://www.iwfa.net/
IWLA	Org	International Warehouse Logistics Association. A for-profit organization assisting 3PL's with warehousing operations. See http://www.iwla.org. Most members of the IWLA are based in North America.
LAA	Org	Logistics Association of Australia. See http://www.logassoc.asn.au/
LESA	Org	Logistics Execution Systems Association. See MHIA.
LGMDA	Org	Lawn & Garden Marketing and Distribution Association. See http://www.lgmda.org/
LLP	Org	Lead Logistics Provider. A master contractor who manages an entire outsourced logistics network for a company. LLPs traits include multi-modal services, advanced technology, international reach, and able to manage complex requirements. AKA as a 4PL.
LMI	Org	Logistics Management Institute. See http://www.lmi.org/
MAD	Org	Michigan Association of Distributors. See http://www.asa.net/

Suggest New Words http://www.IDII.com/addword.htm

© 2004 Industrial Data & Information Inc.

Glossary of Supply Chain Terminology

Word	Type	Definition of Word
Manhattan	Org	Manhattan Associates - provider of supply chain execution solutions worldwide. Manhattan and Red Prairie are considered leaders in the WMS software industry. www.manh.com
Manugistics	Org	A software company providing TMS solutions worldwide. Manugistics and i2 Technologies are considered leaders in the TMS software industry. See http://www.manugistics.com.
MDA	Org	Midwest Distributors Association. See http://www.asaonline.org/
MDA	Org	Music Distributors Association. See http://www.musicdistributors.org/
MHIA	Org	Material Handling Industry of America. See http://www.mhia.org. Within the membership of MHIA, there are divisions. Two of these units are the Logistics Execution Systems Association (LESA) and Order Selection, Staging & Storage Council (OSSSC).
MIS	C Org	Management Information Services, Management Information System. The department or division responsible for computer applications, programmers, developers, quality assurance, documentation of applications, and hot-line support staff. AKA IS and IT.
MIT	Org	Massachusetts Institute of Technology. See http://web.mit.edu/
MTS	Mfg W	Make To Stock. See also CTO and ETO.
NACD	Org	National Association of Container Distributors. See http://www.nacd.net/
NACFAM	Org	National Coalition for Advanced Manufacturing. See http://www.nacfam.org/

Suggest New Words   http://www.IDII.com/addword.htm

© 2004 Industrial Data & Information Inc.

Glossary of Supply Chain Terminology

Word	Type	Definition of Word
NAED	Org	National Distribution of Electrical Distributors. See http://www.naed.org/
NAGASA	Org	National American Graphic Arts Suppliers Association. See http://www.nagasa.org/
NAHAD	Org	National Association of Hose & Accessories Distributors. See http://www.nahad.org/
NAHSA	Org	North American Horticultural Suppliers Association. See http://www.nahsa.org/
NAPM	Org	National Association of Purchasing Management
NAPSA	Org	National Appliance Parts Suppliers Association. See http://www.napsaweb.org/
NARM	Org	National Association of Recording Merchandisers. See http://www.narm.com/
NARS	Org	National Association of Rail Shippers. See http://www.railshippers.com/
NASFT	Org	National Association for the Specialty Food Trade, Inc. See http://www.nasft.org/
NASGW	Org	National Association of Sporting Goods Wholesalers. See http://www.nasgw.org/
NATCD	Org	National Association of Tobacco and Confectionary Distributors. See http://www.natcd.com/
NATD	Org	North American Association of Telecommunications Dealers. See http://www.natd.com/
NAW	Org	National Association of Wholesaler-Distributors. See http://www.naw.org/
NCBAA	Org	National Customs Brokers / Forwarders Association of America. See http://www.ncbfaa.org/
NCC	Org	National Classification Committee.
NFFA	Org	National Frozen Foods Association. See http://www.nffa.org/

Suggest New Words http://www.IDII.com/addword.htm

© 2004 Industrial Data & Information Inc.

Glossary of Supply Chain Terminology

Word	Type	Definition of Word
NFPA	MSDS Org	National Fire Protection Association is an international membership organization which promotes/ improves fire protection and prevention and establishes safeguards against loss of life and property by fire. Best known on the industrial scene for the National Fire Codes (16 volumes of codes, standards, recommended practices and manuals developed and periodically updated by NFPA technical committees). Among these is NFPA 704M, the code for showing hazards of materials as they might be encountered under fire or related emergency conditions, using the familiar diamond-shaped labels or placards with appropriate numbers and symbols. (HC)
NGA	Org	National Grocers Association. See http://www.nationalgrocers.org/
NIOSH	Org	National Institute for Occupational Safety and Health. OSHA and NIOSH are partnering together for guidelines and regulations for worker safety based on research & advisory committee. See http://www.cdc.gov/niosh/homepage.html
NIST	Org	National Institute of Standards Technology. See http://www.nist.gov/
NITL	Org	Institute of National Transport Logistics. See http://www.nitl.ie/
NMFTA	Org	National Motor Freight Traffic Association, Inc. A non-profit organization whose members are primarily LTL motor carriers. The NMFTA maintains the NMFC codes, assigns the SCAC (Standard Carrier Alpha Code) to motor carriers, and more. See http://www.nmfta.org
NMHC	Org	National Materials Handling Center.

Suggest New Words http://www.IDII.com/addword.htm

© 2004 Industrial Data & Information Inc.

Glossary of Supply Chain Terminology

Word	Type	Definition of Word
NPTC	Org	National Private Truck Council. See http://www.nptc.org/
NSPI	Org	National Pool & Spa Institute. See http://www.nspi.com/
NSSEA	Org	National School Supply & Equipment Association. See http://www.nssea.org/
NTP	Org	National Transportation Program, which is part of the DOE of the US Government. See http://www.ntp.doe.gov
NUSA	Org	National Unaffiliated Shippers' Association. See http://www.nusa.net/
OLA	Org	Optical Laboratories Association. See http://www.ola-labs.org/
OMG	Org Std	Object Management Group. See http://www.omg.org/
OPWA	Org	Office Products Wholesalers Association. See http://www.opwa.org/
OSF	Org	Open Software Foundation. See http://www.opengroup.org/
OSHA	MSDS Org	Occupational Safety and Health Administration. See http://www.osha.gov See also DOL and NIOSH, as OSHA collaborates with other government agencies for worker safety guidelines & regulations.
OSSSC	Org	Order Selection, Staging, & Storage Council. See MHIA.
PAC	Org	Political Action Committees.
PASBA	Org	Procurement And Supply-Chain Benchmarking Association. See http://www.pasba.com/
PERA	Org	Production Engine Remanufacturers Association. See http://www.pera.org/
PIDA	Org	Pet Industry Distributors Association. See http://www.pida.org/

Suggest New Words http://www.IDII.com/addword.htm

© 2004 Industrial Data & Information Inc.

Glossary of Supply Chain Terminology

Word	Type	Definition of Word
PMAC	Org	Purchasing Management Association of Canada. See http://www.pmac.ca/
PMI	Org	Project Management Institute. With over 100,000 members worldwide, PMI is the leading professional association in the area of Project Management. See http://www.pmi.org
PSDA	Org	Pacific Southwest Distributors Association. See http://www.asaonline.org/
RICI	Org	Remanufacturing Industries Council International. See http://www.reman.org/
RLEC	Org	Reverse Logistics Executive Council. See http://www.rlec.org/
RosettaNet	Org	RosettaNet. Building standard interfaces for commerce based on XML. Similar to OAG & OAGIS for inter-operability between vendor applications. See http://www.rosettanet.org
SAL	Org	Smart Active Label (SAL) Consortium. See http://www.sal-c.org/
SAP	Org	SAP Inc. A worldwide provider of ERP software. See http://www.sap.com/
SBA	Org	Small Business Administration. See http://www.sba.gov/
SCA	Org	Shipbuilders Council of America. See http://www.shipbuilders.org/
SCC	Org	Supply Chain Council. See http://www.supply-chain.org/
SCL	Org	Canadian Association of Supply Chain & Logistics Management. See http://www.infochain.org/
SCRA	Org	Specialized Carriers and Rigging Association. See http://www.scranet.org/
SEDA	Org	Safety Equipment Distributors Association. See http://www.safetycentral.org/

Suggest New Words   http://www.IDII.com/addword.htm

© 2004 Industrial Data & Information Inc.

Glossary of Supply Chain Terminology

Word	Type	Definition of Word
SEWA	Org	Southeastern Warehouse Association. See http://www.se-warehouseassoc.org/
SHIELD	Org Tr	Shippers for International Electronic Logistics Data. This Org promotes e-commerce best practices and expedites international trade through automation.
SIG	Org	Special Interest Group. This specialized group has a very specific purpose to accomplish and belongs to a larger Org. See http://sigchi.org/
SME	Org	Society of Mechanical Engineers. See http://www.asme.org/
SOLE	Org	International Society of Logistics. Society of practitioners representing commercial, government, and defense. See http://www.sole.org
SSCF	Org	Sanford Supply Chain Forum. See http://www.stanford.edu/group/scforum/
SSWA	Org	Sanitary Suppliers Wholesale Association. See http://www.sswa.com/
STB	Org Tr	Surface Transportation Board. See www.stb.dot.gov/
SWA	Org	Southern Wholesalers Association. See http://www.asaonline.org/
TAC	Org	Transportation Association of Canada. See http://www.tac-atc.ca/
T&B	Org	Tibbett & Britten Group plc. See www.tibbett-britten.com/
TBG	Org	Tibbett & Britten International. See www.tibbett-britten.com/
TCA	Org	Truckload Carriers Association. See http://www.truckload.org/
TCPC	Org	Transportation Consumer Protection Council. See http://www.transportlaw.com/
TIA	Org	Transportation Intermediaries Association. See www.tianet.org

Suggest New Words   http://www.IDII.com/addword.htm

© 2004 Industrial Data & Information Inc.

Glossary of Supply Chain Terminology

Word	Type	Definition of Word
TLI	Org	The Logistics Institute. The Logistics Institute at Georgia Tech (TLI) has pioneering research and education programs in supply chain, transportation, and e-logistics. Good classes & white papers. See http://tli.isye.gatech.edu
TMA	Org	Tooling and Manufacturing Association. See http://www.tmanet.com/
TPC	Org	Transaction Processing Performance Council. See http://www.tpc.org/
TRA	Org	Transportation Research Board. See http://www.nas.edu/trb/
UCC	Org	Uniform Code Council. Founded 1972. A not-for-profit standards development organization. The UCC & the EAN International, administers the EAN UCC, UPC and more. See http://www.uc-council.org
UKWA	Org	United Kingdom Warehousing Association. The Association represents third party logistics companies in the United Kingdom. See http://www.ukwa.org.uk
UN	Org	United Nations. See http://www.un.org/
UPS	Org	United Parcel Service. One of the largest parcel carriers in North America and worldwide. See http://www.ups.com.
USDA	Org	United States Department of Agriculture. See http://www.usda.gov.
USPO	Org	United States Post Office. See http://www.uspo.gov. The USPO is a "carrier" for letters & parcels.

Suggest New Words http://www.IDII.com/addword.htm

© 2004 Industrial Data & Information Inc.

Glossary of Supply Chain Terminology

Word	Type	Definition of Word
W3C	Org	World Wide Web Consortium. W3C has over 400 Member organizations and was created in October 1994. Its mission is to develop common protocols (E.G., SOAP, XHTML, CSS, etc.) that promote the WWW's growth and ensure its interoperability. See http://www.w3c.org
WANE	Org	Wholesalers Association of the Northeast. See http://www.asaonline.org/
WDA	Org	Wholesale Distributors Association. See http://www.asaonline.org/
WECA	Org	Wireless Ethernet Compatibility Alliance – an organization that serves & promotes the wireless industry. See www.wirelessethernet.org.
WERC	Org	Warehousing Education and Research Council. Non-profit association with educational & research focus on warehousing. See http://www.werc.org
WFFSA	Org	Wholesale Florist & Florist Supplier Association. See http://www.wffsa.org/
WHO	Org	World Health Org. See www.who.int/en/
WI-FI	Org	The Wi-Fi Alliance is a nonprofit international association formed in 1999 to certify interoperability of wireless Local Area Network products based on IEEE 802.11 specification. See http://www.wi-fi.org
WSA	Org	Western Suppliers Association. See http://www.asaonline.org/
WSWA	Org	Wine & Spirits Wholesalers Association. See http://www.wswa.org/

Suggest New Words http://www.IDII.com/addword.htm

© 2004 Industrial Data & Information Inc.

Pallet

Terminology
Utilized In
Pallets & Containers
Of Products

Suggest New Words http://www.IDII.com/addword.htm

© 2004 Industrial Data & Information Inc.

Glossary of Supply Chain Terminology

Word	Type	Definition of Word
Annular Nail	Pa	Pallet nail with annular (circular ring) threads rolled onto the shank. (P1)
Banding Notch	Pa	See Strap Slot. (P1)
Bin	Pa W	Four-sided superstructure to be mounted on a pallet base, with or without a cover; also known as a box or container bin pallet. (P1)
Block	Pa	Rectangular, square or cylindrical deck spacer, often identified by its location within the pallet as corner block, end block, edge block, inner block, center or middle block. (P1)
Block Pallet	Pa	A type of pallet with blocks between the pallet decks or beneath the top deck. (P1)
Bottom Deck	Pa	Assembly of deckboards comprising the lower, load-bearing surface of the pallet. (P1)
Butted Deckboard	Pa	An inner deckboard placed tightly against an adjacent lead deckboard during pallet assembly. (P1)
CAD	Pa	Computer-aided-design software that allows the design of the "right" pallet at the best value. See Pallet Design System (PDS). (P1)
Captive Pallet	Pa	A pallet intended for use within the confines of a single facility, system or ownership; not intended to be exchanged. (P1)
Chamfered Deckboards	Pa	Deckboards with edges of one or two faces beveled, either along the full or specified length of board or between the stringers or blocks, allowing easier entry of pallet jack wheels. (P1)

Suggest New Words http://www.IDII.com/addword.htm

© 2004 Industrial Data & Information Inc.

Glossary of Supply Chain Terminology

Word	Type	Definition of Word
CHEP	Org Pa	CHEP founded in 1946 and is a multi-national company known primary for CHEP pallets that it rents. The abbreviation CHEP stands for "Commonwealth Handling Equipment Pool". CHEP is equipment pooling system provider. See http://www.chep.com.
Collar	Pa	Collapsible wooden container or bin, which transforms a pallet into a box. (P1)
Cost-Pass Through	Pa	A cost-share system where the partial cost of a pallet is passed-through from the purchaser to the receiver of the pallet. (P1)
Cost-Per-Trip	Pa	Average cost of pallet use for a single one-way trip. (P1)
Deck	Pa	One or more boards or panels comprising the top or bottom surface of the pallet. (P1)
Deckboard	Pa	Element or component of a pallet deck, oriented perpendicular to the stringer or stringerboard. (P1)
Deckboard Spacing	Pa	Distance between adjacent deckboards. (P1)
Deckboard Span	Pa	Distance between deckboard supports (stringers, stringerboards or blocks). (P1)
Deck Mat	Pa	Assembly of deckboards and stringerboards, forming the deck of a block pallet. (P1)
Deck Opening	Pa	The space between the deckboards of a pallet. (KA)
Deflection	Pa	The amount of deformation or bending in a pallet or pallet component under load. (P1)
Dimensions	Pa	See Pallet Dimensions. (P1)
Double-Face Pallet	Pa	A pallet with top and bottom deckboards extending beyond the edges of the stringers or stringerboards. (P1)

Suggest New Words   http://www.IDII.com/addword.htm

© 2004 Industrial Data & Information Inc.

Glossary of Supply Chain Terminology

Word	Type	Definition of Word
Drive Screw Nail	Pa	Helically (continuous spiral) threaded pallet nail. (P1)
Economic Life	Pa	Output from program that identifies the number of trips the pallet will make, provided it is properly repaired, which maximizes a return on investment. (P1)
Exchange Pallet	Pa	A pallet intended for use among a designated group of shippers and receivers where ownership of the pallet is transferal with the ownership of the unit load; common pool pallet. (P1)
Expendable Pallet	Pa	A pallet intended for a series of handlings during a single trip from shipper to receiver; it is then disposed. See Shipping Pallet. (P1)
Fastener	Pa	A mechanical devise for joining pallet components such as nails, staples, bolts or screws. (P1)
Fastener Shear Index	Pa	Relative measure of shear resistance of the pallet fastener. (P1)
Flush Pallet	Pa	A pallet with deckboards flush with the stringers, stringer-boards, or blocks along the sides of the pallet. (P1)
Fork Entry	Pa	Opening between decks, beneath the top deck or beneath the stringer notch to admit forks. (P1)
Four-Way Block Pallet	Pa	A pallet with openings at both pallet ends and along pallet sides sufficient to admit hand-pallet jacks; full four-way entry pallet. (P1)
Free Span	Pa	The distance between supports in a warehouse rack. (P1)
Hand (Wheel) Jack Opening	Pa	Space provided in the bottom deck to allow pallet jack wheels to bear on the floor. (P1)

Suggest New Words   http://www.IDII.com/addword.htm

© 2004 Industrial Data & Information Inc.

Glossary of Supply Chain Terminology

Word	Type	Definition of Word
Handling	Pa	A single pick-up, movement and set-down of a loaded or empty pallet. (P1)
Hardened-Steel Nail	Pa	Heat-treated and tempered steel pallet nail with a MIBANT angle between 8 and 28 degrees. (P1)
Hardwood	Pa	Wood from broad-leaved species of trees (not necessarily hard or dense). (P1)
Helical nail	Pa	Helically (continuous spiral) threaded pallet nail. See also Drive Screw Nail. (P1)
Inner deckboard	Pa	Any deckboard located between the end deckboards. (P1)
Joint	Pa	Intersection and connection of components, often identified by location within the pallet as the end joint, center joint and corner joint. (P1)
Lateral Collapse	Pa	The failure of pallet joints due to extreme forces. The force occurs in a direction perpendicular to the stringerboard. (KA)
Length	Pa	Refers to the stringer or stringerboard (in block pallets) length; also refers to the first dimension given to describe a pallet, l. (P1)e. (P1), 48" x 40", where 48" is the pallet stringer/stringerboard length. (P1)
Life To First Repair	Pa	Output from PDS program which is equivalent to the number of trips the pallet will last before needing repair. (P1)
Line Load	Pa	The weight of a unit load concentrated along a narrow area across the full length or width of the pallet. (P1)
Load Bearing Surface	Pa	Actual area of material in contact with and supporting a unit load. (P1)
Low Lift	Pa	See Pallet Jack. (KA)

Suggest New Words http://www.IDII.com/addword.htm

© 2004 Industrial Data & Information Inc.

Glossary of Supply Chain Terminology

Word	Type	Definition of Word
Mat	Pa W	A panel of wood, rubber, or other material that is placed on top of unit loads to allow tight strapping of the load without product damage. (KA)
MIBANT	Pa	Morgan Impact Bend Angle Nail Tester.
MIBANT Angle	Pa	The bend angle in a fastener shank when subjected to a MIBANT test. (P1)
MIBANT Test	Pa	Standard impact nail tester used in the pallet and lumber industry as an indication of impact bend resistance of nails or staples. (P1)
Nail	Pa	Fastener made from endless wire by cutting a point and forming a head at the shank end opposite the point. (P1)
Nail Diameter	Pa	The average diameter of a nail shank. (KA)
Nail Head Diameter	Pa	The average diameter of a nail head. (KA)
Nail Length	Pa	The distance measured parallel to the shank from the top of the head to the point of the nail. (KA)
Nail Shank	Pa	The length of the nail, not including the tip or head. (KA)
Non-Reversible Pallet	Pa	A pallet with bottom deckboard configuration different from top deck. (P1)
Notch	Pa	Cutout in lower portion of the stringer to allow entry for the fork tine, usually 9" in length, 1-1/2" in depth. (P1)
Notched stringer	Pa	A stringer with two notches spaced for fork-tine entry, (partial four-way entry). (P1)
Notch Filet	Pa	The curvature at either corner of a pallet notch. (KA)
Notch Height	Pa	The distance between the bottom surface of a pallet stringer and the top of the notch. (KA)

Suggest New Words http://www.IDII.com/addword.htm

© 2004 Industrial Data & Information Inc.

Glossary of Supply Chain Terminology

Word	Type	Definition of Word
NWPCA	Org Pa	National Wooden Pallet and Container Association - A national association with the goal of promoting the de- sign, manufacturer, distribution, recycling and sale of pallets, containers and reels. (P1)
Opening Height	Pa	The vertical distance measured between decks, from the floor to the underside of the top deck, or from the floor to the top of the stringer notch. (P1)
Overall Height	Pa	The vertical distance measured from the floor to the top side of the top deck. (P1)
Overhang	Pa	The distance the deck extends from the outer edge of the stringer or stringerboard; wing; lip. (P1)
Pallet	Pa	A portable, horizontal, rigid platform used as a base for assembling, storing, stacking, handling and transporting goods as a unit load, often equipped with a superstructure. (P1)
Pallet Design System	Pa	Reliability based computer-aided design (CAD) program, for determining the safe load carrying capacity, performance, life and economy of wooden pallets. (P1)
Pallet-Dimensions	Pa	When specifying pallet size, the stringer or stringerboard (block pallet) length is always expressed first; for example, a 48" x 40" pallet has a 48" stringer or stringerboard and 40" deckboards. (P1)
Palletization	Pa W	System for shipping goods on pallets. Permits shipment of multiple units as one large unit. (Cnf)
Palletize	Pa W	To place material on a pallet in a prescribed area. (KA)
Palletizer	Pa W	A type of materials-handling device suing conveyor or robotics to position cubes or bags on a pallet. (KA)

Suggest New Words http://www.IDII.com/addword.htm

© 2004 Industrial Data & Information Inc.

Glossary of Supply Chain Terminology

Word	Type	Definition of Word
Pallet Jack	Pa W	Hand-propelled, wheeled platform, equipped with a lifting device for moving palletized unit loads. AKA low lift or pallet mover. (P1)
Pallet Life	Pa	The period during which the pallet remains useful, expressed in units of time or in the number of one-way trips. (P1)
Panel Deck Pallet	Pa	Pallet constructed with composite or structural panel top deck. (P1)
Partial Four-way Stringer Pallet	Pa	A pallet with notched stringers. (P1)
PDS	Pa	See Pallet Design System.
Post Pallet	Pa	A pallet fitted with posts or blocks between the decks or beneath the top deck. See Block Pallet. (P1)
Racked Across Deckboards	Pa	Output from PDS program describing the maximum load carrying capacity and deflection of a pallet where the rack frame supports the pallet only at the ends of the deckboards. (P1)
Racked Across Stringers	Pa	Output from PDS program describing the maximum load carrying capacity and deflection of a pallet where the rack frame supports the pallet only at the ends of the stringers or stringer boards. (P1)
Recycling	Pa	A pallet, container or reel that has been used, discarded, salvaged, repaired and which passes through a cycle again. (P1)
Rental Pallet	Pa	A pallet owned by a third party, different from the actual pallet user. (P1)
Repair	Pa	To remake in order to use again. (P1)
Returnable Pallet, Reusable Pallet	Pa	A pallet designed to be used for more than one trip. (P1)

Suggest New Words http://www.IDII.com/addword.htm

© 2004 Industrial Data & Information Inc.

Glossary of Supply Chain Terminology

Word	Type	Definition of Word
Shipping Pallet	Pa	Pallet designed to be used for a single one-way trip from shipper to receiver; it is then disposed. See Expendable Pallet. (P1)
Shook	Pa	Cut-to-size pallet parts to be assembled into pallets. (P1)
Shook Grade	Pa	The classification of the quality of pallet parts relative to performance characteristics based on size and distribution of defects, independent of wood species. (P1)
Single-Wing Pallet	Pa	A pallet with the deckboards extending beyond the edges of the stringers or stringer-boards with the bottom deckboards flush (if present). (P1)
Skid	Pa	A pallet having no bottom deck. (P1)
Slave Pallet	Pa	Pallet, platform or single, thick panel used as a support base for a palletized load in rack-storage facilities or production systems. (P1)
Soft Nail	Pa	Pallet nail with a MIBANT angle equal to or greater than 47 degrees. (P1)
Softwood	Pa	Wood from coniferous or needle-bearing species of trees (not necessarily soft or low density). (P1)
Solid Deck Pallet	Pa	A pallet constructed with no spacing between deckboards. (P1)
Spacer	Pa	A pallet component that is located between top and bottom deckboards or beneath a single top deck. A spacer creates the opening that enables forks to be inserted into the pallet. (KA)
Span	Pa	The distance between stringer or block supports. (P1)
Stevedore Pallet	Pa	A pallet designed for use on seaport shipping docks, normally of heavy-duty, double-wing construction. (P1)

Suggest New Words http://www.IDII.com/addword.htm

© 2004 Industrial Data & Information Inc.

Glossary of Supply Chain Terminology

Word	Type	Definition of Word
Stiff-Stock Steel Nail	Pa	Pallet nail made of medium-high carbon steel without heat treatment and tempering with MIBANT angle between 29 and 46 degrees. (P1)
Strap Slot	Pa	Recess or cutout on the upper edge of the stringer or the bottom of the top deckboard to allow tie-down of a unit load to the pallet deck with strapping/banding, also called the banding notch strapping - thin flat bands used to secure load to pallet. (P1)
Stringer	Pa	Continuous, longitudinal, solid or notched beam-component of the pallet used to support deck components, often identified by location as the outside or center stringer. (P1)
Stringerboard	Pa	In block pallets, continuous, solid board member extending for the full length of the pallet perpendicular to deckboard members and placed between deckboards and blocks. (P1)
Stringer Chord	Pa	The upper edge of a notched stringer. (KA)
Stringer Foot	Pa	The lower edge of a notched stringer. (KA)
Take-it-or-leave-it Pallet	Pa	A pallet fitted with fixed cleats on the top deckboards to permit fork truck tines to pass beneath the unit load and remove it from the pallet. (P1)
Tie-Sheets	Pa	Pallet-size pieces of rough cardboard or fiberboard used between tiers to stabilize unitized loads. (KA)
Top Cap	Pa	Panel to be placed on top of a unit load to allow for tight strapping without damaging the unit load. (P1)
Top-deck of the Pallet	Pa	The assembly of deckboards comprising the upper load-carrying surface of the pallet. (P1)
Trip	Pa	Consists of four to six handlings of a pallet. (P1)

Suggest New Words   http://www.IDII.com/addword.htm

© 2004 Industrial Data & Information Inc.

Glossary of Supply Chain Terminology

Word	Type	Definition of Word
Two-way Entry Pallet	Pa	A pallet with unnotched solid stringers allowing entry only from the ends. (P1)
Unit Load	Pa W	Assembly of goods on a pallet for handling, moving, storing and stacking as a single entity. (P1)
Warehouse Pallet	Pa	A double-face multiple trip returnable pallet intended for general warehouse use. (P1)
Wing	Pa	Overhang of deckboard end from outside edge of stringer or stringer. (P1)

Suggest New Words http://www.IDII.com/addword.htm

© 2004 Industrial Data & Information Inc.

Purchasing

Terms & Words
Utilized In
Purchasing &
Procurement of Goods

Glossary of Supply Chain Terminology

Word	Type	Definition of Word
Alpha Factor	Pu	In exponential smoothing forecasting, the smoothing constant applied to the most recent forecast error. (KA)
Demand Forecasting	Pu	Determining predictions of a product's future usage using a judgmental approach, a relational approach, or a time series-approach. (KA)
EOQ	Pu	Economic Order Quantity. A method to calculate quantity to purchase based on historical needs. One method of many.
Exponential Smoothing	Pu	A forecasting model that uses weighted sum of all prior observations. Weights decline exponentially with the age of the operations. (KA)
Forecast	Pu	An estimate or prediction of future demand. (KA)
Forecast Error	Pu	The difference between actual and forecasted demand. (KA)
LT	Pu	See Lead Time.
MOQ	Pu	Minimum Order Quantity required by the supplier.
Overshipment	Pu W	A shipment containing more than originally ordered. (KA)
PO	B Pu	Purchase Order. Document sent to a vendor indicating goods or services to be purchased at a specific price and terms for indicated quantities. See EDI 850.

Suggest New Words http://www.IDII.com/addword.htm

© 2004 Industrial Data & Information Inc.

Glossary of Supply Chain Terminology

Word	Type	Definition of Word
Procurement	Pu	The first phase of the supply process. Procurement includes all activities from the selection of vendors to the purchase and transport of raw materials to the point of manufacture, reprocessing, or repackaging. (KA)
PSCM	Pu	Procurement In Supply Chain Management (EM)
Purchase Requisition	Pu	An intracompany form that instructs the firm's purchasing agent to acquire merchandise or supplies. (KA)
Purchasing Agent	Pu	The individual authorized to purchase goods and services for a company. (KA)
Purchasing Costs	Pu	All costs associated with the purchasing activities. These costs include personnel, office supplies, and communication charges. (KA)
Re-Order Point	Pu	When on-hand quantity of the specified product decreases to this re-order point or less, it is time to reorder.
REQ	Pu	Purchase Requisition.
ROP	Pu	Re-order Point. When on-hand quantity of the specified product decreases to this re-order point or less, it is time to reorder.
Safety Stock	Pu	A minimum quantity of stock carried in inventory in addition to the forecasted customer requirements. Safety stock is meant to provide sufficient stock for emergencies, unanticipated demand, or unforeseen delays. (KA)
SAMI	Pu	Supplier Assistance in Managing Inventories (EM)
Seasonal Index	Pu	A number used to adjust forecasts and optimal inventory calculations for seasonal fluctuations. (KA)

Suggest New Words http://www.IDII.com/addword.htm

© 2004 Industrial Data & Information Inc.

Glossary of Supply Chain Terminology

Word	Type	Definition of Word
Seasonal Inventory	Pu W	Inventory held to meet seasonal demand. (KA)
Single Sourcing	Pu	The process of using a single supplier for many goods or services. Price competition is reduced, but interdependence should be increased. (KA)
SMI	Pu	Supplier Managed Inventories (EM)
Vendor Compliance	Pu	A program in which retailers require vendors to provide detailed information, either on bar coded packing slips or via EDI for inbound receipts. (KA)
WFC	Pu	Weeks Forward Coverage.

Suggest New Words http://www.IDII.com/addword.htm

© 2004 Industrial Data & Information Inc.

Standards

Established
& Proposed Standards

Glossary of Supply Chain Terminology

Word	Type	Definition of Word
2D Barcode	Std	Barcode Symbology.
2 of 5	Std	Barcode Symbology.
3 of 9	Std	Barcode with a variable length for alphanumeric codes. Commonly utilized in industry. No check digits utilized in 3 of 9 barcodes. Also known as Code 39.
16K	Std	Barcode Symbology.
802.11a	C Std	Wireless network standard that operates at 5 GHz with a maximum throughput of 54 Mbps. Range is 80 feet and supplies 12 separate non-overlapping channels. Good for high performance in a small, limited area. Incompatible with existing 802.11b and 802.11g equipment.
802.11b	C Std	Wireless network standard that operates at the crowded 2.4 GHz with a maximum throughput of 11 Mbps. Range is 328 feet and supplies 3 separate non-overlapping channels. Equipment is low cost and available. Compatible with existing 802.11g equipment.
802.11g	C Std	Wireless network standard that operates at the crowded 2.4 GHz with a maximum throughput of 54 Mbps. Range is 328 feet and supplies 3 separate non-overlapping channels. Interoperates with existing 802.11b equipment. Good for larger campuses.
AES	C Std	Advanced Encryption Standard. See also DES & WEP.
ANSI	Std	American National Standards Institute. ANSI provides information and conformity assessments to the US standards development community and industry. See http://www.ansi.org

Suggest New Words http://www.IDII.com/addword.htm

© 2004 Industrial Data & Information Inc.

Glossary of Supply Chain Terminology

Word	Type	Definition of Word
ASCII	Std	American Standards Code for Information Interchange.
Bookland	Std	Barcode symbology used on books to encode ISBN on books.
Code 11	Std	Barcode Symbology.
Code 16K	Std	Barcode Symbology.
Code 39	Std	Barcode Symbology.
Code 128	Std	Barcode symbology with variable length for alphanumeric codes. More compact and higher validation than Code 39.
Codeabar	Std	Barcode symbology that has been used by Federal Express, libraries, and other niche markets.
CORBA	Std	Common Object Request Broker Architecture.
CPFR	Std	Collaborative Planning, Forecasting and Replenishment. VICS Group sponsored this standard to optimize a manufacturers production and delivery of their product. See http://www.cpfr.org
DataMatrix	Std	2D barcode symbology.
DES	C Std	Data Encryption Standard. See also AES & WEP.
EAN	Std	European Article Numbering.
EAN	Std	International Article Numbering Association.
EDIFACT	Std EDI	EDI for Administration, Commerce and Transport. EDIFACT is n EDI standard for international usage and was developed by the United Nations.
EPC	Std	Electronic Product Code (ePC) is 28 digits. Retailers and standard groups are moving to this ePC code. See also GTIN and RSS.

Suggest New Words   http://www.IDII.com/addword.htm

© 2004 Industrial Data & Information Inc.

Glossary of Supply Chain Terminology

Word	Type	Definition of Word
FIPS	Std	Federal Information Processing Standards.
GLN	Std	Global Location Number. A GLN identifies the entities of the organization. A GLN can identify corporate locations (headquarters, divisions, departments, warehouses, shipping locations, buyers' offices, bill-to offices...). See EAN-UCC for more information.
GTIN	Std	Global Trade Item Number. A 14-digit code representing a product, which most retailers and manufacturers are using or implementing now. The GTIN can be used to synchronize data via UCCnet's GLOBAL registry. See also UCC, ePC, RSS.
HIBC	Std	Health Industry Bar Code.
Interleaved 2 of 5	Std	Barcode symbology.
ISO 9000	Std	International Standards Organization Standards for Quality Systems.
ISSN	Std	Barcode symbology used to encode ISSN numbers for periodicals outside of North America.
JAN	Std	Barcode symbology used in Japan, which is similar to EAN.
Ladder	Std	A barcode printed vertically so that the bars appear to be a ladder.
Maxicode	Std	2D barcode symbology used by the United Parcel Service (UPS).
Modified Plessey Code	Std	Barcode symbology.

Suggest New Words http://www.IDII.com/addword.htm

© 2004 Industrial Data & Information Inc.

Glossary of Supply Chain Terminology

Word	Type	Definition of Word
MSI Plessey	Std	Barcode symbology.
NDC	Std	National Drug Code. See also GTIN, RSS, and UCC.
NIACAP	Std	National Information Assurance Certification and Accreditation Process.
OAG	Std	Open Applications Group. Building standard OAGIS via XML for business applications. Focus is on standard data formats and interfacing for inter-operability between external vendor applications. See http://www.openapplications.org
OMG	Org Std	Object Management Group. See http://www.omg.org/
PDF417	Std	2D stacked barcode symbology where one may encode a large amount of information within the PDF417 barcode.
Picket Fence	Std	A barcode printed horizontally so that the bars look like a picket fence.
Postnet	Std	Barcode symbology used on mail for the US Post Office to encode zip codes (postal codes).
Quiet Zone	Std	A blank area around the printed barcode this is required. This quiet zone size is small and varies upon the barcode symbology.
RSS	Std	Reduced Space Symbology code from the UCC is 14 digits. A technique used to conserve space in the barcode for UCC/EAN. See also GTIN and ePC.

Suggest New Words http://www.IDII.com/addword.htm

© 2004 Industrial Data & Information Inc.

Glossary of Supply Chain Terminology

Word	Type	Definition of Word
SCAC	Std Tr	Standard Carrier Alpha Code is a unique code to identify transportation companies. The Standard Carrier Alpha Code is the recognized transportation company identification code used in the American National Standards Institute (ANSI) Accredited Standards Committee (ASC) X12 and United Nations EDIFACT approved electronic data interchange (EDI) transaction sets. The NMFTA assigns SCAC codes for all motor carriers. See http://www.nmfta.org
SCC, SCC-14	Std	Shipping Container Code consisting of 14 digits. This UCC SCC-14 code is frequently put on a pallet or container label. The first two digits are the UCC application identifier. The next digit is the packaging level indicator. The next seven digits is the manufacturer identification number. The next five digits represent the manufacturer assigned product number. The last digit is a modulus 10 check digit. See also UCC 128.
SIC	Std	Standard Industry Classification. Coding system to classify different industries including manufacturers, wholesalers, retailers, and service providers.
SITC	Std	See Standard International Trade Classification.
SOAP	Std	Simple Object Access Protocol. Data transfer of business data types using XML and SOAP. See also XML, Cobra, OMG, and OASIS.
SQL	C Std	Standard Query Language. A common programming language to query and maintain (add, update, and delete) data. SQL varies slightly from database vendor to vendor.

Suggest New Words http://www.IDII.com/addword.htm

© 2004 Industrial Data & Information Inc.

Glossary of Supply Chain Terminology

Word	Type	Definition of Word
SSCC	Std	Serial Shipping Container Code. 18-digit license plate code for a container. UCC & EAN are instrumental in this specification.
Standard International Trade Classification	Std	A numerical code developed by the United Nations and adopted by U.S. airlines as the basis for identifying commodities moving in air freight. (Cnf)
UCC	Std	See Uniform Commercial Code.
UCS	Std	Uniform Container Symbol. See UCC 128.
UDDI	Std	Universal Description, Discovery, and Integration. An online directory to facilitate business-to-business electronic commerce. See http://www.uddi.org for specifications, registering your company, or searching for services or products.
Universal Commercial Code	Std	Abbreviated UCC. The law used in every state except Louisiana to govern business transactions. (KA)
Universal Product Code	Std	Abbreviated UPC. See UPC-A, UPC-E, UPCC, and the UCC for specifications.
UPC	Std	Universal Product Code. See UPC-A, UPC-E, UPCC, and the UCC for specifications.
UPC-A	Std	Barcode with a fixed length of 12 digits for point-of-sale retail product usage. The twelve digits break into: First digit is a product indicator. Next 5 digits is the manufacturer id. Next 5 digits is the product id number. Last digit is a check digit.
UPCC	Std	Uniform Product Carton Code.
UPC-E	Std	Barcode symbology by EAN/UPC that encodes a UCC-12 identification number in 6 digits.
VICS	Std	Voluntary Inter-Industry Commerce on standards. Multiple projects including the CPFR. See http://www.vics.org
VPN	C Std	Virtual Private Network.

Suggest New Words http://www.IDII.com/addword.htm

© 2004 Industrial Data & Information Inc.

Glossary of Supply Chain Terminology

Word	Type	Definition of Word
WEP	C Std	Wired Equivalent Privacy. An encryption standard. For those using wireless 802.11, it is best to use an encryption to keep wireless messages secured, such as WEP, AES, DES, or MAC.
WSDL	C Std	Web Services Definition Language. See http://www.w3.org/TR/wsdl for more details.
X12	Std	ANSI EDI standards committee has produced the EDI X12 standards for business documents.
XBRL	Std	Stands for eXtensible Business Reporting Language. An open specification that uses XML-based data tags to describe financial data in business reports & databases. This makes pulling and pushing of data faster and more fluid!
XML	Std	Data format specification called eXtensible Markup Language. Standard groups have defined business transaction standards in XML. See OAG, BizTalk, RosettaNet.
XQL	Std	XML Query Language. Ability to query a database via an XML based query.

Suggest New Words http://www.IDII.com/addword.htm

© 2004 Industrial Data & Information Inc.

Transportation

Terms & Words
Utilized In The Transportation Industry

Glossary of Supply Chain Terminology

Word	Type	Definition of Word
Abandonment	Tr	1. Proceeding where a carrier seeks authorization to stop service over all or part of its route/line, or to give up ownership/control of cargo or vessel. Must be approved by the ICC in the case of motor or rail proceedings. 2. Shipper or consignee relinquishes damaged freight carrier or refuses to accept delivery. 3. The act or relinquishing title to damaged or lost property to claim a total loss. (Cnf)
ABI	Tr	Automated Broker Interface.
Absolute Liability	Tr	Condition in which a carrier is responsible for all liability and is not protected by normal exemptions found in a bill of lading or common law liability. (Cnf)
Acceptance	Tr	1. Acknowledged receipt by consignee of a shipment, terminating the common carrier contract. 2. A promise to pay, usually evidenced by inscribing across the face of the bill "accepted," followed by the date, place payable, and acceptor's signature. (Cnf)
Accessorial Charges	Tr	Charges for supplementary services and privileges (Accessorial Services) provided in connection with line-haul transportation of goods. These charges are not included in the freight charge and usually take the form of a flat fee. Some examples are: pickup/delivery, in-transit privileges, demurrage, switching, loading/unloading. (Cnf)
Accessorial Services	Tr	A service rendered by a carrier in addition to regular transportation service. (Cnf)

Suggest New Words http://www.IDII.com/addword.htm

© 2004 Industrial Data & Information Inc.

Glossary of Supply Chain Terminology

Word	Type	Definition of Word
ADA	Gov Tr	Airline Deregulation Act of 1978. (WA)
Ad Valorem	Tr	A Latin phrase meaning "according to value." Freight rates set at a certain fixed percentage of the value of articles are known as ad valorem rates. (Cnf)
Advanced Charge	Tr	Freight or charge on a shipment that is advanced by one transportation company to another, or to the shipper, to be collected from the consignee. (Cnf)
Advice of Shipment	Tr	Notice to local or foreign buyer that shipment has occurred, with packing and routing details. A copy of invoice usually is enclosed, and sometimes a copy of the bill of lading. (Cnf)
AES	Tr	Automated Export System.
AESTIR	Tr	Automated Export System Trade Interface Requirements
AGTM	Tr	Automated Global Trade Management. A computer & transportation term for software for advanced regulatory content management and trade automation services.
Air Waybill	Tr	A non-negotiable bill of lading that covers both domestic and international flights transporting goods to a specified destination. (WA) Abbreviated AWB.
All-Commodity Rate	Tr	Usually a carload/truckload rate that applies to multiple shipments that move at one time in one vehicle from the consignor to one consignee. An all-commodity rate is established based on actual transportation cost rather than "value of service." (Cnf)
Allowance	Tr	Deduction from the weight or value of goods. Allowed if a carrier fails to provide necessary equipment and that equipment is furnished by the shipper. (Cnf)

Suggest New Words http://www.IDII.com/addword.htm

© 2004 Industrial Data & Information Inc.

Glossary of Supply Chain Terminology

Word	Type	Definition of Word
Alongside	Tr	Point of delivery beside a vessel; statement designating where the title to goods passes from one party to another. (Cnf)
Alternate Routing	Tr	Routing that is less desirable than the normal, but results in identical terms. (Cnf)
Alternative Terms	Tr	The terms that a rail carrier offers in place of its full liability terms. The alternative terms are cheaper and carry less than a full limitation of liability. (WA)
AMS	Tr	Automated Manifest System. US Customs Automated Manifest System (AMS). TMS and SMS solutions can become AMS certified. Being AMS certified enables users of that software solution to submit advance cargo manifest information via the Automated Manifest System (AMS). This ensures that shipments won't face penalties, unnecessary delays and disruption to their supply chain.
Arbitrary	Tr	1) Charge in addition to regular freight charge to compensate for unusual local conditions. 2) Fixed amount accepted by a carrier when dividing joint rates. (Cnf)
Arrival Notice	Tr	A notice, furnished to consignee, of the arrival of freight. (Cnf)
Astray Freight	Tr	Freight that has been separated from its freight bill. (Cnf)
ATA	Tr	Actual Time of Arrival. When the carrier's vehicle arrives at the agreed upon physical location.
ATD	Tr	Actual Time of Departure. When the carrier's vehicle departs from the agreed upon physical location.
ATRI	Org Tr	American Transportation Research Institute, formerly the ATA Foundation. See http://www.atri-online.org/

Suggest New Words http://www.IDII.com/addword.htm

© 2004 Industrial Data & Information Inc.

Glossary of Supply Chain Terminology

Word	Type	Definition of Word
Auditing	Tr	Determining the correct transportation charges due the carrier, either before or after payment. Auditing involves checking the freight bill for extension errors, correct rate, commodity classification and description, weight, density, etc. Also called freight audit. (WA)
Authority	Tr	Operating rights granted a motor carrier by the ICC. (Cnf)
Authorized Carrier	Tr	A person or company authorized by the ICC to transport goods as a common or contract carrier. (Cnf)
AWB	Tr	See Air Waybill.
Back Haul	Tr	1. Return transportation movement, usually at less revenue than the original move. 2. Movement in the direction of lighter traffic flow when traffic generally is heavier in the opposite direction. 3. To move a shipment back over part of a route already traveled. (Cnf)
Barrell Truck, Barrell Wheeler	Tr	A dolly-like hand truck designed specifically to move drums or barrels. (Cnf)
Basing Point	Tr	Geographic point to which transportation rates are set so that rates to adjacent points can be constructed by adding to/deducting from the basing point rate. (Cnf)

Suggest New Words http://www.IDII.com/addword.htm

© 2004 Industrial Data & Information Inc.

Glossary of Supply Chain Terminology

Word	Type	Definition of Word
Bill of Lading	Tr	Bills of lading are contracts between the owner of the goods and the carrier. There are two types: A straight bill of lading in not negotiable. A negotiable or shipper's order bill of lading can be bought, sold or traded while goods are in transit and is used for letter of credits transactions. The customer usually needs the original or a copy as proof of ownership to take possession after payment for the goods. (WA) Abbreviated BOL, BL, B/L.
Billed Weight	Tr	The weight shown on a freight bill. (Cnf)
Birdyback	Tr	Moving highway freight by air. (Cnf)
BL, B/L	Tr	See Bill of Lading.
Blind Side	Tr	Right side of truck and trailer. (Cnf)
Blocking	Tr	Wood or metal supports to keep shipments in place in trailers. (Cnf)
Bob Tail	Tr	Tractor operating without a trailer. (Cnf)
Bogey	Tr	A two-axle assembly at the rear of some trailers or tractors. Also called a tandem axle. (Cnf)
BOL	Tr	See Bill of Lading.
Bottom Freight	Tr	Heavy freight that must be loaded on the trailer floor and not on top of other merchandise. (Cnf)
Box	Tr	Slang term for a trailer or container for ocean carriers. Slang term for a truck transmission. (Cnf)
Bracing	Tr	See Blocking.
Break Bulk	Tr	The separation of a consolidated bulk load into smaller individual shipments for delivery to the ultimate consignee. The freight may be moved intact inside the trailer, or it may be interchanged and rehandled to connecting carriers. (WA)
Breakdown Time	Tr	A type of penalty pay that is incurred when equipment breaks down. (Cnf)

Suggest New Words http://www.IDII.com/addword.htm

© 2004 Industrial Data & Information Inc.

Glossary of Supply Chain Terminology

Word	Type	Definition of Word
Broker	Tr	1. An agent who arranges interstate movement of goods by other carriers. 2. An arranger of exempt loads for owner-operators and/or carriers. 3. One who arranges the buying/selling of goods for a commission. 4. A person who leases owned equipment to a carrier. (Cnf)
Bulk Carrier	Tr	Vessel that carries bulk commodities such as petroleum, grain, or ore, which are not packaged, bundled, bottled, or otherwise packed. (Cnf)
Bulk Freight	Tr	Freight not in packages or containers such as wheat, petroleum, etc. (Cnf)
Bulkhead	Tr	1) An upright wall in a trailer or railcar that separates and stabilizes a load. 2) A cargo restraining partition in a vehicle or vessel. (Cnf)
Buyer's Right to Route	Tr	When a seller does not pay freight charges, the purchaser has a right to designate the route for shipment. Seller is responsible for following the buyer's instructions. Complete routing is permitted for rail shipments, but only for the first carrier in motor shipments. (Cnf)
Cab	Tr	Driver's compartment of a truck or tractor. (Cnf)
Cabatoge	Tr	A federal law that requires coastal and intercoastal traffic to be carrier in U.S. built and registered ships. (WA)
Cable Seal	Tr	A heavy steel cable used to secure closed trailer doors. It can only be removed with heavy-duty cable cutters. (Cnf)

Suggest New Words   http://www.IDII.com/addword.htm

© 2004 Industrial Data & Information Inc.

Glossary of Supply Chain Terminology

Word	Type	Definition of Word
Cab-Over	Tr	Truck or tractor with a substantial part of the driving cab located over the engine. (Cnf)
CAF	Tr	Cost and Freight. Quoted price includes the cost of transportation only. Buyer is responsible for loss & damage of shipment and may insure it at their expense. See also CIF
Capacity Load	Tr	1) A trailer loaded to the legal weight limit. 2) A trailer loaded so that no additional piece of freight, equal to the size of the largest piece tendered, will fit into the trailer. (Cnf)
CAPSTAN	Tr	Computer-Aided Planned Stowage and Networking system.
Carbatoge	Tr	A federal law that requires coastal and intercoastal traffic to be carrier in U.S. built and registered ships. (WA)
Carload	Tr	1) Quantity of freight required to fill a railcar. 2) Specified quantity necessary to qualify a shipment for a carload rate. Abbreviated as C/L and CL. (Cnf)
Carmack Amendment	Gov Tr	A federal statute that codifies the common law principle that a carrier is liable for the full value of goods lost, damaged or delayed while in its possession unless the shipper agrees in writing to a lower rate in return for a lower limitation of liability offered by the carrier. (WA)
Carrier	Tr	The commercial entity responsible for deliveries of shipment to the customer. Carriers may specialize in small packages, air cargo, ocean cargo, rail cargo, LTL, or full truck load (TL). See also common carrier, contract carrier, authorized carrier.

Suggest New Words http://www.IDII.com/addword.htm

© 2004 Industrial Data & Information Inc.

Glossary of Supply Chain Terminology

Word	Type	Definition of Word
Carrier Liability	Tr	A common carrier is liable for all shipment loss, damage, or delay with the exception of that caused by an act of God, act of a public enemy, act of a public authority, act or omission of the shipper, or the inherent vice or nature of the goods. (WA)
Cart	Tr	A four-wheeled platform used to move several pieces of freight across the dock at one time. (Cnf)
Cartage	Tr	1) The charge for pickup/delivery of goods. 2) The act of moving goods (usually short distances). (Cnf)
Carte Blanche	Tr	Full discretionary power, unlimited authority. (WA)
Case Mark	Tr	Information shown on the outside of a shipping carton, including destination and contents. (Cnf)
Cash Before Delivery	Tr	Seller assumes no risk and extends no credit because payment is received before shipment. Abbreviated CBD. (Cnf)
Cash On Delivery	Tr	Buyer pays carrier the price of goods when they are delivered; seller assumes risk of purchaser refusing to accept goods. Abbreviated COD. (Cnf)
CASM	Tr	Carrier Assessment Selection and Management (EM)
CBD	Tr	See Cash Before Delivery.
CCRA	Org Tr	Canada Customs and Revenue Agency.
CDL	Tr	Commercial Driver's License. In the US, the FMCSA is monitoring security for those applying for a CDL to transport hazardous materials.

Suggest New Words http://www.IDII.com/addword.htm

© 2004 Industrial Data & Information Inc.

Glossary of Supply Chain Terminology

Word	Type	Definition of Word
C&F	Tr	Cost and Freight. Quoted price includes the cost of transportation only. Buyer is responsible for loss & damage of shipment and may insure it at their expense. See also CIF
CFS	Tr	See Container Freight Station.
Chock	Tr	A wooden, metal, or rubber wedge used to block the wheels of a trailer at the dock. Also used in trailers to keep floor freight from shifting. (Cnf)
CIF	Tr	Cost, Insurance, and Freight. Quoted price includes cost of the goods, insurance of the goods, and freight charges. See also C&F.
CKD	Tr	Complete Knock Down logistics. I.E., Logistics provider would transport complete cars or motors broken down in parts to be assembled elsewhere.
C/L, CL	Tr	See Carload
Claim	Tr	1. Demand on transportation company for payment due to loss/damage of freight during transit. 2. Demand on transportation company for refund on overcharge. 3. Demand by an individual/company to recover for loss under insurance policy. (Cnf)
Claimant	Tr	The party filing a claim or suit. (WA)

Suggest New Words http://www.IDII.com/addword.htm

© 2004 Industrial Data & Information Inc.

Glossary of Supply Chain Terminology

Word	Type	Definition of Word
Classifications	Tr	An alphabetical listing of commodities, assigning a class or rating into which the commodity is placed, with the truckload minimum weight necessary for application of a truckload rate; used primarily to determine less-than-truckload rates based on class rates. (WA)
Class Rate	Tr	The rate charged for commodities grouped according to similar shipping characteristics. Class Rate applies to numbered/lettered groups/classes of articles contained in the territorial rating column in classification schedules. (Cnf)
COD	Tr	See Cash On Delivery.
Codification	Tr	The process of compiling, arranging, and systematizing the laws of a given jurisdiction into an ordered code. (WA)
COFC	Tr	See Container on a Flat Car. (KA)
Collectively Made Rates	Tr	Class rates established collectively by carrier members of regional rate bureaus and rate conferences under a grant of immunity from federal antitrust laws. (WA)
Collect Shipment	Tr	Shipment where collection of freight charges/advances is made by delivering carrier from the consignee/receiver. (Cnf)
Co-Load	Tr	Two shipments from different terminals combined to ship as one load. (Cnf)
Combination	Tr	Truck or tractor coupled to one or more trailers (including semi-trailers). (Cnf)
Commercial Invoice	B Tr	Itemized list issued by seller/exporter in foreign trade showing quantity, quality, description of goods, price, terms of sale, marks/numbers, weight, full name/address of purchaser, and date. (Cnf)
Commodity, Exempt	Tr	One that may be transported in interstate commerce without operating authority or published rates. (Cnf)

Suggest New Words http://www.IDII.com/addword.htm

© 2004 Industrial Data & Information Inc.

Glossary of Supply Chain Terminology

Word	Type	Definition of Word
Commodity Rate	Tr	1. A special (usually lower) rate for specific types of goods (usually exempt commodities). 2. A rate lower than class rates, established to cover the movement of a specific customer's freight or for a specific group of customers. (Cnf)
Common Carriage	Tr	Carriage performed by a common carrier that is not under contract with the shipper. (WA)
Common Carrier	Tr	A for-hire carrier that holds itself out to serve the general public. (WA)
Concealed Damage	Tr W	When goods in an apparently undamaged container are damaged. (Cnf)
Concurrence	Tr	Document signed by carrier and filed with the ICC. Verifies carrier participates in rates published in a tariff by a given agent. (Cnf)
Connecting Carrier	Tr	A carrier that originates or completes transportation of a shipment, but does not haul it the entire distance from origin to destination. (Cnf)
Consign	Tr	Send goods to a purchaser or an agent to sell. (Cnf)
Consignee	Tr	The person who receives goods that were shipped.
Consignor	Tr	The sender of a freight shipment. This is often termed the Seller or Vendor.
Consolidating	Tr	Combining small shipments to obtain reduced freight rated for higher volume. (KA)
Consolidation Point	Tr	The point at which small shipments are combined and loaded for reshipment. (KA)
Consolidator	Tr	A company that specializes in providing consolidation services to shippers. (KA)

Suggest New Words http://www.IDII.com/addword.htm

© 2004 Industrial Data & Information Inc.

Glossary of Supply Chain Terminology

Word	Type	Definition of Word
Constructive Notice	Tr	A rule of law binding the shipper to the terms and conditions contained in a law or a filed tariff. (WA)
Container	Tr W	Anything in which articles are packed. A standardized box used to transport merchandise particularly in international commerce. Marine containers are typically 8'X8' with length of 10', 30', or 40'. These containers may be transloaded from rail cars or ships onto a truck frame and delivered to their final destination. (KA)
Container Freight Station	Tr	A physical location by the carrier for cargo to be packed or unpacked by the carrier.
Container-ization	Tr W	1. Using box-like device to store, protect, and handle a number of packages as a unit of transit. 2. Shipping system based on large cargo-carrying containers that can be interchanged between trucks, trains, and ships without re-handling contents. (Cnf)
Container on a Flat Car	Tr	Abbreviatd COFC. A trailer without chassis or intermodal container shipped on a railroad flat car. (KA)
Contract	Tr	Legally defined as an offer, acceptance and bargained for consideration, a contract may be oral or in writing, if there is a valid offer, acceptance, and something of value exchanged between the two parties. (WA)
Contract Carriage	Tr	Transportation under a bilateral contract between a shipper and a carrier for a continuing period of time. (WA)

Suggest New Words   http://www.IDII.com/addword.htm

© 2004 Industrial Data & Information Inc.

Glossary of Supply Chain Terminology

Word	Type	Definition of Word
Contract Carrier	Tr	Any carrier engaged in interstate transportation of persons/property by motor vehicle on a for-hire basis, but under continuing contract with one or a limited number of customers to meet specific needs of each customer. Contract Carriers must receive an authorization permit from the ICC. (Cnf)
Contract of Adhesion	Tr	A standard-form contract prepared by one party, to be signed by the party in a weaker position. (WA)
Contract Rates	Tr	Rates that are part of a total contract negotiated between shipper and a carrier. (Cnf)
Conventional	Tr	Tractor with the engine in front of the cab. (Cnf)
Cost, Insurance, and Freight	Tr	The basis for quotation by seller that indicates seller will pay insurance and freight charges to destination only. (Cnf)
Credit Rules	Tr	Tariff rules whereby the carrier agrees to deliver goods without receiving payment prior to delivery. The rules stipulate when payment may be paid without incurring a penalty, and may impose a penalty for late payments if the rules are published in accordance with FMCSA regulations. (WA)
Cross Dock	Tr W	Transfer of freight from one trailer to another at a terminal. (Cnf)
CSI	Tr	U.S. Customs Container Security Initiative. Places US Customs inspectors overseas to identify high-risk shipments. See www.customs.ustreas.gov See also C-TPAT, FAST, and HSA.

Suggest New Words http://www.IDII.com/addword.htm

© 2004 Industrial Data & Information Inc.

Glossary of Supply Chain Terminology

Word	Type	Definition of Word
C-TPAT	Tr	Customs-Trade Partnership Against Terrorism. Requires importers, carriers and custom brokers to document security procedures. Accepted companies qualify for expedited processing and exemption from physical inspections. See www.customs.ustreas.gov. See also TSA, CSI, FAST, and HSA.
Cube Loading	Tr W	The process of loading merchandise onto pallets or other unit-loading techniques to allow several unit loads to be stacked for transportation. (KA)
Cube Rate	Tr	A rate based on trailer space instead of weight. Used for light, bulky loads. (Cnf)
Cube Utilization	Tr W	The percentage of space occupied compared to the space available. (KA)
CUSDEC	Tr	Customs Declaration Message. An EDI (UN/EDIFACT) transaction to send information from one party to another.
Customhouse Broker	Tr	A person or firm licensed to enter and clear goods through Customs. (WA)
Customs Bond	Tr	A contract between a principal, usually an importer, and a surety, which is obtained to insure performance of an obligation imposed by law or regulation. The bond covers potential loss of duties, taxes and penalties for specific types of transactions. (WA)
Customs Broker	Tr	A specialist in customs procedures that processes the entry and clearance of goods into the country for a fee.
Customs Tariff	Tr	A schedule of charges assessed by the government on imports/exports. (Cnf)
Damage Claim	Tr	Request by a shipper or consignee for reimbursement from a carrier for damage to a shipment. (KA)

Suggest New Words http://www.IDII.com/addword.htm

© 2004 Industrial Data & Information Inc.

Glossary of Supply Chain Terminology

Word	Type	Definition of Word
Dead Axle	Tr	Non-powered rear axle on tandem truck or tractor (also called "tag axle"). (Cnf)
Deadhead	Tr	1) A trailer moving empty. 2) A shipment moving without charges. 3) A ride-along driver. (Cnf)
Dead Weight Tonnage	Tr	Estimated number of tons of cargo a vessel can carry when loaded to maximum depth. (Cnf)
Declared Value	Tr	1) Assumed value of shipment unless shipper declares higher value. 2) Stating lower value on a shipment to get a lower rate. (Cnf)
Deductible	Tr	An amount of loss accepted by the insured as a condition for a reduction in the insurance premium or carrier's rate. (WA)
Deferred Rebate	Tr	Carrier returns a portion of freight charges to shipper. In exchange, shipper gives all/most shipments to carrier over specified period, usually six months. Rebate payment is deferred for similar period. (Cnf)
Delivery Receipt	Tr	A carrier-prepared form that is signed by the consignee at the time of delivery. See Proof of Delivery (KA)
Delivery Window	Tr	A period during which a delivery (or deliveries) must be made. (KA)
Demurrage	Tr	1) Detention of a ship, freight car, or other cargo conveyance during loading or unloading beyond the scheduled time of departure. 2) Compensation paid for such detention.
Density	Tr W	The weight of an article per cubic foot. (Cnf)
Detention	Tr	A charge made for a vehicle held by, or for, consignor or consignee for loading, unloading or for forwarding directions. (Cnf)

Suggest New Words http://www.IDII.com/addword.htm

© 2004 Industrial Data & Information Inc.

Glossary of Supply Chain Terminology

Word	Type	Definition of Word
Differential	Tr	Amount added to/deducted from base rate to make rate to/from some other point or via another route. (Cnf)
Discount	Tr	A reduction from the full amount of rate and charges offered by the carrier. (WA)
Discrimination	Tr	Difference in rates not justified by costs. (Cnf)
Dispatch	Tr	The process of combining a driver, a tractor, a trailer, and a load. (KA)
Diversion	Tr	A change made in consignee, destination, or shipment route while in transit. (Cnf)
DOA	Tr	Dead On Arrival. Imperfect delivery.
Dock	Tr W	The floor or platform where trucks load and unload. (Cnf)
Dolly	Tr W	A non-motorized, two-wheeled hand truck for moving freight around the dock. (Cnf)
DOT	Tr	See Department of Transportation.
Drayage	Tr	Transporting freight by truck, primarily in local cartage. (Cnf)
Drive Axle	Tr	The axle(s) which are connected to the engine by a drive shaft and power the vehicle. Also called "power axle". (Cnf)
Driver Collect	Tr	Refers to a shipment for which the driver must collect freight charges at the time of delivery. (Cnf)
DSRC	Tr	Dedicated Short-Range Communications. See ASTM and ITS.
Dual Rate System	Tr	An international water carrier pricing system in which a shipper signing an exclusive use agreement with the conference pays a rate 10 to 15 percent lower than non-signing shippers pay for an identical shipment. (WA)
DUB	Tr	A 28-foot trailer designed to be pulled two or three at a time by one tractor. Also known as pup or doubles. (Cnf)

Suggest New Words http://www.IDII.com/addword.htm

© 2004 Industrial Data & Information Inc.

Glossary of Supply Chain Terminology

Word	Type	Definition of Word
Dunnage	Tr	Term used for cardboard, empty pallets, plywood, foam rubber, air bags, or other items used to cushion or protect freight while in transit. (Cnf)
ECCN	Tr	Export Control Classification Number
EIC	Tr	Export Information Code
EMC	Tr	Export Management Company. A firm that serves as an export agent, paid by a fee or retainer basis.
End-Of-The-Line-Terminal	Tr	A local terminal that handles the pick-up and delivery of the customer's freight (as opposed to a consolidation center). Also referred to as a "satellite" or "group" terminal. (Cnf)
EOL	Tr	See End-Of-The-Line-Terminal.
Estoppel	Tr	The doctrine of equitable estoppel means that he who by his language or conduct leads another to do what he would not otherwise have done, shall not subject such person to loss or injury by disappointing the expectations upon which he acted. (WA)
ETA	Tr	Estimated Time of Arrival.
ETC	Tr	See Export Trading Company.
ETD	Tr	Estimated Time of Departure.
Exception	Tr	1) A shortage, overage, or damage to a shipment. 2) A notation of such conditions on a freight bill, bill of lading or unloading check sheet. (Cnf)
Exculpatory Clause	Tr	A contractual provision relieving a party from any liability resulting from a negligent or wrongful act. (WA)
Exempt Carrier	Tr	For-hire motor carrier exempt from ICC economic regulation. (Cnf)
Exempt Circular	Tr	A railroad's tariff applying on commodities exempt from government controls. (WA)

Suggest New Words http://www.IDII.com/addword.htm

© 2004 Industrial Data & Information Inc.

Glossary of Supply Chain Terminology

Word	Type	Definition of Word
Expediting	Tr	Moving shipments through regular channels at an accelerated rate. (Cnf)
Export Broker	Tr	A firm that charges fees to assists foreign buyers with domestic manufacturers.
Export Declaration	Tr	Document declaring goods to be exported which is required by the government.
Export Letter of Credit	Tr	See Letter of Credit.
Export Trading Company	Tr	A firm that purchases goods from domestic manufacturers & sources and sells them to foreign markets. The ETC may work on commission rather than purchasing the goods.
F/A	Tr	Free Astray.
FAK	Tr	Freight All Kinds. An agreed upon freight rate for all types of products from a facility.
FAS	Tr	Freight Alongside Ship.
FAS	Tr	See Free Along Side.
FAST	Tr	Participants in C-TPAT, working with Canada's Partners in Protection (PIP) receives unique identifiers that make them eligible for expedited processing at the US-Canada border. Carriers and Drivers must be pre-approved. www.customs.ustreas.gov See also C-TPAT, CSI, and HSA.
FB	Tr	See Freight Bill.
FCL	Tr	Full Container Load.
Feeder	Tr	In intermodal moves, a pickup/delivery vehicle or ship. (Cnf)
FF	Tr	See Freight Forwarder.
Filed Tariffs	Tr	A carrier's tariff filed with a government agency pursuant to a statute or regulation. (WA)

Suggest New Words http://www.IDII.com/addword.htm

© 2004 Industrial Data & Information Inc.

Glossary of Supply Chain Terminology

Word	Type	Definition of Word
FIRMS	Tr	Facilities Information and Resource Management System code. Identifies the U.S. Customs Service facility where merchandise is located.
Fishy Back	Tr	Transporting motor carrier trailers and containers by ship. (Cnf)
FIT	B Tr	Fabrication in Transit.
Flatbed	Tr	A trailer without sides used for hauling machinery, steel beams, and other lengthy or bulky items. (WA)
Flatcar	Tr	A railcar without sides, used for hauling machinery, steel, etc. (WA)
FMCSA	Org Tr	Federal Motor Carrier Safety Administration. US government agency responsible for overseeing motor carrier safety & guidelines.
FMS	Tr	Freight Management Systems. See TMS.
FOB	Tr	Freight On Board.
FOB	Tr	See Free On Board.
FOB Destination	Tr	Freight costs paid to the destination point, title transfers at destination. (Cnf)
FOB Factory	Tr	Title to goods and transportation responsibility transfers from seller to factory. (Cnf)
FOB Origin	Tr	Free on board up to delivery to the carrier. The consignor retains liability for merchandise until it is picked up by the carrier. The consignee is considered to have title to the goods beginning at the time the carrier accepts the shipment. Typically the consignee is responsible for selecting the carrier and for all charges relating to transportation. (KA)
FOB Shipping Point	Tr	Title to merchandise passes to the consignee at the point when the goods are delivered to the transportation provider. (KA)

Suggest New Words http://www.IDII.com/addword.htm

© 2004 Industrial Data & Information Inc.

Glossary of Supply Chain Terminology

Word	Type	Definition of Word
FOB Vessel	Tr	Title/transportation costs transfer after goods are delivered on vessel. All export taxes/costs involved in overseas shipments are assessed to the buyer. (Cnf)
Force Majeure	Tr	Condition in contract that relieves either party from obligations where major unforeseen events prevent compliance with provisions of agreement. (Cnf)
Foreign Sales Agent	Tr	A firm or person that works to represent & sell your goods in a foreign country. AKA Foreign Sales Representative.
Foreign Trade Zones	B Tr	Goods subject to duty may be brought into such zones duty-free for transshipment/storage/minor manipulation/sorting. Duty must be paid when/if goods are brought from a zone into any part of the U.S. (Cnf)
Forwarding Agent	Tr	A firm specializing in shipping goods abroad. Payments made for insurance and other expenses are charged to the foreign buyer. (Cnf)
Free Alongside	Tr	Selling term in international trade. Selling party quotes price including delivery of goods alongside overseas vessel at exporting port. Abbreviated FAS. (Cnf)
Free-Astry	Tr	A shipment that is mis-routed or unloaded at the wrong terminal and is billed and forwarded to the correct terminal free of charge. (Cnf)
Free On Board	Tr	Loaded aboard carrier's vehicle at point where responsibility for risk/expense passes from seller to buyer. Abbreviated FOB (Cnf)

Suggest New Words http://www.IDII.com/addword.htm

© 2004 Industrial Data & Information Inc.

Glossary of Supply Chain Terminology

Word	Type	Definition of Word
Free Time	Tr	The period freight will be held before storage charges are applied. The period allowed for the owner to accept delivery before storage charges begin to accrue. (Cnf)
Freight	Tr	In North America, freight refers to shipments over 150 lb. to be shipped via LTL, TL, Ocean, Rail, or Air freight. Freight is in contrast to "parcel" shipments, for small packages, under 150 lb.
Freight Audit	Tr	See Auditing.
Freight Bill	Tr	Carrier's invoice for payment of transportation charges. Abbreviated FB. See also Bill Of Lading.
Freight-Broker (Property)	Tr	A person or company responsible for arranging the transportation of goods between points in interstate commerce by motor carrier. As compensation for arranging transport, the broker receives a commission. (KA)
Freight Consolidation	Tr	The merging of shipments from several manufacturers for transport to the same destination under a single bill of lading. This method of shipping provides freight savings for all parties involved. (KA)
Freight Forwarder	Tr	A firm or person that assists with export and other documentation. May assist in the packing and shipping of these goods.
Freight Prepaid	Tr	Notation marked on the face of the bill of lading indicating that the freight bill has been prepaid by the shipper, or that the carrier has extended credit to the shipper. (WA)
FTC	Tr	Freight To Customers.
FTL	Tr	Full Truck Load or Full Trailer Load.
FTZ	B Tr	See Foreign Trade Zones.

Suggest New Words http://www.IDII.com/addword.htm

© 2004 Industrial Data & Information Inc.

Glossary of Supply Chain Terminology

Word	Type	Definition of Word
FVP	Tr	Full value protection. Shippers can opt to pay an additional charge for full replacement value of lost or damaged goods from the carrier.
Gateway	Tr	The point at which freight is interchanged/interlined between carriers or at which a carrier joins two operating authorities provision of through service. (Cnf)
GATT	Tr	General Agreement on Tariffs & Trade.
GPS	Tr	Global Positioning System. Provides accurate latitude and longitude of a position by a series of satellites and receivers to compute the position.
Grandfather Clause	Tr	A provision that enabled motor carriers engaged in lawful operations before the passage of a statute that regulated an industry for the first time, granting applicants a license to continue operations without proving a public need for those services. (referred to as " public convenience and necessity") (WA)
Gross Weight	Tr	Weight of the shipment including packing material. See also Net Weight.
Hazard Class	Tr	A code indicating type of hazardous material. Normally utilized for carriers that indicate which of their services allow certain hazard classes. E.G. UPS 1 Day AIR services do NOT allow aerosols or combustible products.

Suggest New Words   http://www.IDII.com/addword.htm

© 2004 Industrial Data & Information Inc.

Glossary of Supply Chain Terminology

Word	Type	Definition of Word
HazMat	Tr W	Hazardous Material. Product is classified hazardous by a government agency or a carrier. Special handling is required on some hazardous items, which may include segregation, separation wall, temperature control, and/or limitations on how the product may be shipped. See also MSDS and Hazard Class.
HMF	Tr	Harbor Maintenance Fees
Hot Load	Tr	A rush shipment. (KA)
House Air Waybill	Tr	An air waybill that an air freight forwarder issues. All house air waybills are covered by a master air waybill, which represents all the shipments tendered by an air freight forwarder to an airline. (WA)
HSA	Tr	Homeland Security Act. This US Government Act created the Department of Homeland Security (DHS) and brings the Coast Guard, the Transportation Security Administration (TSA) and US Customs under the DHS. http://thomas.loc.gov. See also C-TPAT, CSI, FAST, and TSA.
HTS	Tr	Harmonized Tariff Schedule
ICPA	Org Tr	International Compliance Professionals Association (ICPA) is an organization for trade compliance professionals. See http://www.int-comp.org/
IE	Tr	Immediate Export
ILS	Tr	Integrated Logistics Support.

Suggest New Words http://www.IDII.com/addword.htm

© 2004 Industrial Data & Information Inc.

Glossary of Supply Chain Terminology

Word	Type	Definition of Word
Inadvertence Clause	Tr	A rule in motor carrier tariffs whereby the carrier declines to accept specified commodities valued over a specific amount. However, the rule states that if the carrier "inadvertently" accepts such an item, the carrier's liability for loss or damage will automatically be limited to the stated dollar amount (such as $5.00 per pound, for instance). (WA)
In Bond	Tr	Storage of goods in custody of government/bonded warehouse or carrier from whom goods can be taken only upon payment of taxes/duties to appropriate government agency. (Cnf) See Bonded Warehouse.
Inbound Logistics	Tr	The portion of logistics operations that involves the movement of materials from source to the production plant or warehouse. (KA)
Incorporation By Reference	Tr	Contract term referring to the document or set of documents that the contract refers to but is not included within the four corners of the document. When incorporation by reference applies, the item referred to is considered by courts to be a legally binding part of the contract. (WA)
Incoterms	Tr	Maintained by the International Chamber of Commerce (ICC), this codification of terms is used in foreign trade contracts to define which parties incur the costs and at what specific point the costs are incurred. (WA)

Glossary of Supply Chain Terminology

Word	Type	Definition of Word
Insurance	Tr	Insurance is defined as a contract pursuant to which one party, the insurer, undertakes for compensation, the premium, or the risk of financial loss to another, the insured, against the loss or liability arising from a contingent or unknown event. See Hahn v. Oregon Physicians Serv. et al., 689 F.2d 840, 842 (9th Cir. 1982). (WA)
Inter Alia	Tr	A latin phrase meaning "among other things". (WA)
Interchange	Tr	Passing freight from one carrier to another between lines. (Cnf)
Interline Freight	Tr	Freight moving from origin to destination over two or more transportation lines. (Cnf)
Intermediary	Tr	The term "intermediary" refers to third parties that play a role in arranging for, or providing transportation by carriers for shippers. Intermediaries take the form of truck brokers, foreign, surface, and air freight forwarders, intermodal companies, and third party logistics providers. (WA)
Intermodal Transportation	Tr	Using more than one mode to deliver shipments. For example, rail or ocean vessel carriage of tractor-trailer containers. (Cnf)
Interstate	Tr	The transportation of persons or property between states; in the course of the movement, the shipment crosses a sate boundary. (WA)
IT	Tr	Immediate Transportation
ITL	C Tr	International Trade Logistics. Software for managing international shipment & documentation. Some of the leading ITL software solutions are Vastera, G-Log.

Suggest New Words http://www.IDII.com/addword.htm

© 2004 Industrial Data & Information Inc.

Word	Type	Definition of Word
ITS	Tr	Intelligent Transportation System. ITS systems are in use throughout the United States to promote drive-through toll collection and highway pre-clearance for truckers. Similar technology is in use at the border between NAFTA countries to promote fast and efficient customs checks. In the USA, the FCC's 1999 decision to create a frequency solely for ITS at the 5.9 GHz spectrum provided a new approach the market could endorse, along with allowing for Internet access (data push-pull at ITS access points). The standard is based on IEEE standard 802.11a, which is currently used in private networks. For Road Access & ITS, the standard's group for ITS slightly altered the standard, calling it 802.11/RA (for Road Access.) See also DSRC and ASTM.
Joint and Several Liability	Tr	Liability that may be apportioned either among two or more parties or to only one or a few select members of the group, at the adversary's discretion; thus each liable party is not individually responsible for the entire obligation. (WA)
Joint Rate	Tr	Agreed upon by two or more carriers, published in a single tariff, and applying between point on line of one and point on line of another. May include one or more intermediate carriers in route. (Cnf)
Kickback	Tr	Rebate given by a transportation firm. (KA)

Suggest New Words http://www.IDII.com/addword.htm

© 2004 Industrial Data & Information Inc.

Glossary of Supply Chain Terminology

Word	Type	Definition of Word
Latent Defects	Tr	Faults which are not readily apparent through normal diligence. The carrier is not responsible for latent defects. (KA)
Lay Time	Tr	Downtime during which a ship is being loaded or unloaded and for which there is no demurrage charge. (KA)
Legal Weight	Tr	The weight of the goods and the interior packaging but not the container. (KA)
LC	Tr	See Letter of Credit.
LCL	Tr	Less than Container Load. International shipments utilize containers. When the shipment is less than a full container, the freight will be calculated as LCL. See Less Than Container Load.
L&D	Tr	Loss and damage. This term is usually applied when loss or damage is discovered at delivery time. The term "located loss or damage" is used when the damage occurs at an *identifiable* time. (KA)
LDOR	Tr	Local Delivery Operations for Replenishment (EM)
LES	Tr W	Logistics Execution Systems. Involves transportation management (TMS) and warehouse management (WMS) software. AKA Supply Chain Execution (SCE) or Logistics Resource Management (LRM).
Lessee	Tr	Party or company with legal possession/control of vehicle (with/without driver), or other equipment owned by another under terms of lease agreement. (Cnf)
Lessor	Tr	Party or company granting legal use of vehicle (with/without driver), or other equipment to another party under terms of lease agreement. (Cnf)

Suggest New Words   http://www.IDII.com/addword.htm

© 2004 Industrial Data & Information Inc.

Glossary of Supply Chain Terminology

Word	Type	Definition of Word
Less Than Container Load	Tr	A quantity of freight less than that required for the application of container rate. International shipments utilize containers. When the shipment is less than a full container, the freight will be calculated as LCL.
Less Than Truck Load	Tr	A quantity of freight less than that required for the application of truckload rate.
Letter of Credit	Tr	Buyer's bank issues these document that guarantees payment to vendor, with conditions that goods are shipped properly and proper documents are presented. Abbreviated L/C, LC, or LOC.
Lien	Tr	Legal device whereby the lien holder is able to secure the owner's goods until payment or resell them. The lien must be properly filed and attached or the lien holder may be found guilty of conversion. (WA)
Linehaul	Tr	Movement of freight between cities that are usually more than 1,000 miles apart, not including pickup and delivery service. (Cnf)
Load Factor	Tr	1) The weight in pounds loaded onto a trailer. 2) A term used loosely to describe the "compactness" or over "good usage" of trailer space. (Cnf)
Loading Allowance	Tr	A tariff provision which provides an allowance, usually a fixed sum per hundredweight, to a shipper for loading a carrier's trailer. (Cnf)
LOC	Tr	See Letter of Credit.

Suggest New Words   http://www.IDII.com/addword.htm

© 2004 Industrial Data & Information Inc.

Glossary of Supply Chain Terminology

Word	Type	Definition of Word
Logistics	Tr W	Logistics plans, implements, and controls the efficient, effective forward and reverse flow and storage of goods, services, and related information between the point of origin and the point of consumption in order to meet customers' requirements. (Cnf)
LOI	B Tr	Letter of Intent
LO/LO	Tr	Lift On, Lift Off vessel. The method or process in which cargo is loaded unto and unloaded from an ocean vessel.
Longshoreman	Tr	A person who loads and unloads marine vessels. (KA)
LRM	Tr W	Logistics Resource Management. Software that includes modules with transportation management, warehouse/inventory management, and contracting/rate negotiations. AKA as Logistics Execution System (LES) and Supply Chain Execution (SCE).
LTL	Tr	See Less Than Truckload.
Lumping	Tr	Force a truck driver to hire those other than himself to unload his lading. (WA)
Manifest	Tr	A control document used to list the contents (individual shipments) during loading and from which the contents are checked during unloading. (Cnf)
Marine Lien	Tr	A lien on a vessel given to secure the claim of a creditor who provided maritime services to the vessel or who suffered an injury as a result of the vessel's use. (WA)
Market Dominance	Tr	A railroad's control of traffic for which a shipper has no alternative transportation. (WA)

Suggest New Words http://www.IDII.com/addword.htm

© 2004 Industrial Data & Information Inc.

Glossary of Supply Chain Terminology

Word	Type	Definition of Word
Master/Agent Principle	Tr	General tort principle generally referring to the ability of the agent, in most cases an employee, to make liable the master, employer, for action that the agent undertook. (WA)
Master BOL	Tr	See BOL. A master BOL is a BOL document listing all items to be delivered. Usually, the master BOL is on top of the separate BOLS to form a "package" for the driver or carrier.
Master Trip Lease	Tr	An agreement between either a regulated carrier and a private carrier or two regulated carriers. The master trip lease covers multiple loads handled by the two parties for a period of 30 days or more. The master trip lease outlines the responsibilities of each party regarding licensing, insurance, fuel permits, DOT safety compliance, maintenance, vehicle and driver identification and rates of pay. (KA)
Matsui Amendment	Gov Tr	49 U.S.C. § 10502 (c)(1) requiring railroads to maintain full value rates on exempt traffic if they offer reduced rates. (WA)
Meet	Tr	A process by which two drivers going in opposite directions, exchange loads in order to keep the freight moving while allowing the drivers to return to their home locations. (KA)
Meltdown	Tr	The failure of an entire system to handle functions like informatics, routing, etc. (WA)
Merger	Tr	The combination of two or more carriers into one company that will own, manage, and operate the properties that previously operated separately. (WA)

Suggest New Words http://www.IDII.com/addword.htm

© 2004 Industrial Data & Information Inc.

Glossary of Supply Chain Terminology

Word	Type	Definition of Word
MIB	Tr	Master-In-Bond. Paperless Master-In-Bond reporting by carriers that are MIB-approved.
Milk Run	Tr	A pick-up route with multiple stops. (KA)
Mix	Tr	Refers to the combination of light, medium and heavy density freight. (Cnf)
Mode	Tr	A method of transporting materials such as truck, rail, air, ocean barge, and intermodal. (KA)
MOS	Tr	Method of Shipment (EM)
MOT	Tr	Method of Transportation
Motor Carrier Act of 1935	Tr	Act of US Congress bringing motor common and contract carriers under ICC jurisdiction. (Cnf)
Motor Carrier Act of 1980	Tr	Act of Congress that deregulated for-hire-trucking. (Cnf)
Mule	Tr	A small tractor used to move two-axle dollies. The mule is also called a yard tractor. (KA)
Multimodal	Tr	See Intermodal Transportation.
National Motor Freight Classification	Tr	A publication for motor carriers containing rules, descriptions, and ratings on all commodities moving in commerce. The NMFTA maintains these codes. Carriers will base their freight rates on the NMFC codes (unless one utilizes FAK).
Net Weight	Tr	Weight of shipment not including packing materials and container. See also Gross Weight.
NMFC	Tr	See National Motor Freight Classification.
NOIBN	Tr	Not otherwise indexed by name. A rate classification that is similar to FAK but not as broad. It covers all commodities that are not specifically described in the tariff. See Freight of all Kinds. (KA)

Suggest New Words   http://www.IDII.com/addword.htm

© 2004 Industrial Data & Information Inc.

Glossary of Supply Chain Terminology

Word	Type	Definition of Word
Non-Vessel Operation Common Carrier	Tr	Consolidates and issues their own BOL for shipment on a vessel. Does not own nor operate the vessel. Abbreviated NVOOC.
Notice of Arrival	Tr	On arrival of freight at destination, notice is sent promptly to the consignee showing number of packages, description of articles, route, rate, weight, etc. (Cnf)
NVOCC	Tr	Non-Vessel Operation Common Carrier. Consolidates and issues their own BOL for shipment on a vessel. Does not own nor operate the vessel.
OBC	Tr	On Board Computer. Many trucks have OBC to communicate with the company's dispatch and position tracking software.
Off-Route Points	Tr	Points located off regular route highways of line-haul carriers, generally served only on irregular schedules. (Cnf)
Operating Authority	Tr	Routes, points, and types of traffic that may be served by carrier. Authority is granted by state or federal regulatory agencies. (Cnf)
Operating Ratio	Tr	Comparison of carrier's operating expenses with gross receipts; income divided by expenses. Abbreviated as OR and O/R. (Cnf)
OR, O/R	Tr	See Operating Ratio.
OSD	Tr W	See Over, Short, and Damaged.
Outbound Logistics	Tr	The portion of logistics that primarily involves the movement of materials and products from a company's production plant or storage warehouses. (KA)
Overage	Tr W	Freight in excess over quantity believed to have been shipped, or more than quantity shown on shipping document. (Cnf)
Overcharge	Tr	Where a shipper has been overcharged on its freight bill from the carrier. (WA)

Suggest New Words   http://www.IDII.com/addword.htm

© 2004 Industrial Data & Information Inc.

Glossary of Supply Chain Terminology

Word	Type	Definition of Word
Over, Short, and Damaged	Tr W	A report issued at the warehouse when goods are damaged. Used to file a claim with the carrier. (Cnf)
Owner-Operator	Tr	Drivers who own or operate their own trucks. May lease rig/driver to another carrier. (Cnf)
Owner's Risk	Tr	When owner of goods remains responsible during shipping, relieving carrier of part of risk. (Cnf)
Parcel	Tr	In North America, parcel refers to shipments less than 150 lbs to be shipped via a small parcel carrier such as UPS, Fedex, Airborne, and USPO. Freight refers to shipments over 150 lbs to be shipped via LTL, TL, Ocean, Rail, or Air freight
Payload	Tr	1) Carried cargo. 2) The net weight of the cargo. (Cnf)
P&D	Tr	Pickup and Delivery.
Per Diem	Tr	Latin term meaning "by the day." Daily charge for use of railcars. (Cnf)
Perishable Freight	Tr	Commodities subject to rapid deterioration or decay, which require special protective services such as refrigeration or heating. (Cnf)
Permit	Tr	Authority granted to contract carriers and forwarders by the ICC to operate in interstate commerce. (Cnf)
Piggyback	Tr	A form of intermodal transportation where trailers/containers are carried on railcars. (Cnf)
POD	Tr	See Proof of Delivery.
POE	Tr	See Port of Entry.
Point of Origin	Tr	The station at which a shipment is received from the shipper by a transportation line. (Cnf)

Suggest New Words http://www.IDII.com/addword.htm

© 2004 Industrial Data & Information Inc.

Glossary of Supply Chain Terminology

Word	Type	Definition of Word
Pooling Agreement	Tr	The dividing of revenue/business among two or more carriers in accordance with previous contracts/agreements. (Cnf)
Port of Entry	Tr	A port at which foreign goods are admitted into the receiving country. (WA) Abbreviated POE.
Privacy of Contract	Tr	Contract term meaning the parties legally bound and able to recover under the contract. (WA)
Private Carrier	Tr	A company that provides its own transportation, either through leased or owned equipment. Private carriers are allowed to transport goods for subsidiaries and to backhaul products from non-affiliated companies. (KA)
Pro Number	Tr W	A unique number assigned to a freight shipment by the carrier. This Pro Number is utilized on the freight bill and bill of ladings.
Proof Of Delivery	Tr	Copy of waybill signed by consignee at time of delivery as receipt. Abbreviated POD. (Cnf)
Proportional Weight	Tr	Lower than normal rate on segment of through movement to encourage traffic or capture competitive traffic. May be percentage of standard rate or flat rate that is lower between given points. (Cnf)
Pro Se	Tr	On one's own behalf. (WA)
PU&D, PUD	Tr	Pick-Up and Delivery. P&D.
Quid Pro Quo	Tr	The act of exchanging one thing for another. (WA)
Quotas	Gov Tr	Many governments have established quotas of limiting imports by class of goods or country of origin. Sometimes importing countries require issuance of licenses before U.S. companies may ship to them. (Cnf)

Suggest New Words http://www.IDII.com/addword.htm

© 2004 Industrial Data & Information Inc.

Glossary of Supply Chain Terminology

Word	Type	Definition of Word
Rate	Tr	Established shipping charge for movement of goods. In interstate transportation, price/rate is approved by ICC. Intrastate prices are approved by public service commission or similar body. (Cnf) See also FAK.
Rate Basis	Tr	Formula of specific factors/elements that control making of rate. (Cnf)
Rate Bureaus	Tr	A carrier group that assembles to establish joint rates, to divide joint revenues and claim liabilities, and to publish tariffs. (WA)
Rate Files	Tr	The collection of published transportation prices or tariffs. (KA)
Rate War	Tr	When carriers cut rates in an effort to secure tonnage. Can occur in all commodities. (Cnf)
Rebate	Tr	Unlawful practice in which carrier returns part of transportation charge to shipper. Done to encourage shipper to use same carrier again. (Cnf)
Reciprocity	Tr	1. An exchange of rights. In motor transportation, may involve granting equal rights to vehicles of several states in which reciprocity agreements are in effect. 2. To give preference in buying to vendors who are customers of buying company. (Cnf)
Reconsignment	Tr	1) A change (usually requested by the shipper) in the routing, or destination, of a shipment that is already in transit but does not require a new freight bill. 2) A fee for the latter. (See also DIVERSION) (Cnf)
Reefer	Tr	A refrigerated flat bed trailer.
Refrigerated Container	Tr	An insulated container that provides temperature-controlled environment to protect perishable materials. (KA)

Suggest New Words http://www.IDII.com/addword.htm

© 2004 Industrial Data & Information Inc.

Glossary of Supply Chain Terminology

Word	Type	Definition of Word
Released Rates	Tr	Rates based upon the shipment's value. The maximum carrier liability for damage is less than the full value, and in return the carrier offers a lower rate. (WA)
Restricted Articles	Tr	Commodities that can be handled only under certain specific conditions. (Cnf)
Return Receipt	Tr	A form sent to the shipper, after a consignee has received a shipment, that indicates delivery has been made. (KA)
RM	B Tr	Revenue Management. Determination of total cost, labor, and how to determine pricing for revenue analysis. Being done especially with Cargo Revenue Management.
Roll-On / Roll-Off	Tr	A feature in specially constructed vessels permitting road vehicles to drive on/off a vessel in loading/discharging ports. (Cnf)
RO/RO	Tr	See Roll-On / Roll-Off.
Route	Tr	1. Course/direction that shipment moves. 2. To designate course/direction shipment shall move. 3. Carrier(s) with junction points over which shipment moves. (Cnf)
Routing	Tr	1. Process of determining how shipment will move between origin and destination. Routing information includes designation of carrier(s) involved, actual route of carrier, and estimated time en route. 2. Right of shipper to determine carriers, routes and points for transfer on TL and CL shipments. (Cnf)
RTLS	Tr W	Real Time Locating System. Usually in reference to a WMS or YMS ability to move-track-locate inventory.

Suggest New Words http://www.IDII.com/addword.htm

© 2004 Industrial Data & Information Inc.

Glossary of Supply Chain Terminology

Word	Type	Definition of Word
Satellite	Tr	1. A local servicing terminal. Also called "city," "group," or "end-of-the-line" terminals. 2. An accounting designation for a specific service area; not a regular freight terminal location. (Cnf)
SCAC	Std Tr	Standard Carrier Alpha Code is a unique code to identify transportation companies. The Standard Carrier Alpha Code is the recognized transportation company identification code used in the American National Standards Institute (ANSI) Accredited Standards Committee (ASC) X12 and United Nations EDIFACT approved electronic data interchange (EDI) transaction sets. The NMFTA assigns SCAC codes for all motor carriers. See http://www.nmfta.org
Seal	Tr	A lockable numbered metal strip applied to the door of a retail car, truck, or container. A broken seal indicates that the door has been opened. (KA)
SED	Tr	See Shipper's Export Declaration.
Semi	Tr	Slang term for semi-trailer. Also used loosely in referring to tractor-trailer combination. See Semi-Trailer. (Cnf)
Semi-Trailer	Tr	1. Trailer equipped with rear wheels only. The front of the trailer is supported by landing legs when not hooked to power. 2. Generally used to refer to a full size (45 to 48-foot) trailer; as opposed to a doubles trailer. (Cnf)

Suggest New Words http://www.IDII.com/addword.htm

© 2004 Industrial Data & Information Inc.

Glossary of Supply Chain Terminology

Word	Type	Definition of Word
Service of Process Agents	Tr	The person that an entity engaged in business must make available to everyone for the purpose of serving notice that a legal action has been instituted against that entity (WA)
SHIELD	Org Tr	Shippers for International Electronic Logistics Data. This organization promotes e-commerce best practices and expedites international trade through automation.
Shipment	Tr W	1. Lot of freight tendered to carrier by one consignee at one place at one time for delivery to one consignee at one place on one bill of lading. 2. Goods/merchandise in one or more containers, pieces, or parcels for transportation from one shipper to single destination. (Cnf)
Shipper	Tr	The party who tenders the goods to the carrier for movement. Can also be known as the consignee. (WA)
Shipper's Certificate	Tr	Form filled out and presented by shipper to outbound carrier at transit point, together with instructions and inbound carrier's freight bill, asking for reshipping privilege and transit rate on commodity previously brought into transit point. (Cnf)
Shipper's Export Declaration	Tr	Form required by Treasury Department and completed by shipper showing value, weight, consignee, destination, etc., of export shipments, as well as Schedule B identification. Abbreviated SED. (Cnf)
Shipper's Load and Count	Tr	Indicates that the contents of a trailer were loaded and counted by the shipper, the trailer was sealed by the shipper, and the carrier did not observe the loading process. Abbreviated SL&C. (Cnf)

Suggest New Words http://www.IDII.com/addword.htm

© 2004 Industrial Data & Information Inc.

Glossary of Supply Chain Terminology

Word	Type	Definition of Word
Short Shipment	Tr	Piece of freight missing from shipment as stipulated by documents on hand. (Cnf)
Shrink Wrap	Tr W	A plastic wrap used by shippers to secure cartons on a pallet. (Cnf)
Sleeper	Tr	Tractor with a sleeping compartment in the cab. (Cnf)
Stacker	Tr	An individual who loads the freight onto a truck or unloads it. (KA)
Statutory Notice	Gov Tr	Length of time required by law for carriers to give notice of changes in tariffs, rates, rules and regulations — usually 30 days unless otherwise permitted by authority from ICC or other regulatory body. (Cnf)
STB	Org Tr	Surface Transportation Board.
Stevedore	Tr	Person in charge of loading/unloading ships. (Cnf)
Stevedoring	Tr	The unloading of a vessel when in port. (WA)
Straight Bill of Lading	Tr	Non-negotiable document provides that shipment is to be delivered direct to party whose name is shown as consignee. Carrier does not require its surrender upon delivery except when needed to identify consignee. See Bill of Lading. (Cnf)
Stripping	Tr	Emptying truck of cargo, and arranging shipments by destination. (Cnf)
Stuffing	Tr	Slang term for loading cargo container. (Cnf)
Subject Matter Jurisdiction	Tr	The subject matter jurisdiction deals with the power of the court to adjudicate the matter before it and cannot be conferred by agreement. (WA)
Subrogation	Tr	The substitution of a person or entity (usually an insurance company) to the rights of another person or entity (usually the insured) regarding a claim or a debt which the former person paid for the later. (WA)

Suggest New Words http://www.IDII.com/addword.htm

© 2004 Industrial Data & Information Inc.

Glossary of Supply Chain Terminology

Word	Type	Definition of Word
Subrogee	Tr	The person or entity that succeeds the rights of another through subrogation. (WA)
Tanktainer	Tr	Tank built into standard container frame and used to transport liquids. (Cnf)
Tare Weight	Tr	1. Weight of container and material used for packing. 2. In transportation terms, it is the weight of the car/truck, exclusive of contents. (Cnf)
Tariff	Tr	A publication that contains a carrier's rates, accessorial charges, and rules. (WA)
TDPU	Tr	Time Definite Pick/Up (EM)
T&E	Tr	Transportation & Exportation
Tender	Tr	An offer of goods for transportation by a shipper, or an offer of delivery by a carrier. (Cnf)
Through Bill of Lading	Tr	This document covers goods moving from point of origin to final destination, even if transfers are made to different carriers in transit. (Cnf) See Bill of Lading.
Title	Tr	Document that confers on holder right of ownership/possession/transfer of merchandise specified, e.g., bills of lading and warehouse receipts. (Cnf)
TL	Tr	See Truck Load.
TMS	Tr	Transportation Management System. A 'broad' term that includes any type of software dealing with logistics & transportation.
TOFC	Tr	Trailer On Flat Car. A trailer on a railroad flat car.
Tonnage	Tr	1) The carrying capacity of a ship/vessel. 2) The tax/duty paid on such capacity. 3) The weight a ship will carry, expressed in tons. (Cnf)
TPS	Tr	Transportation Planning and Scheduling.

Suggest New Words http://www.IDII.com/addword.htm

© 2004 Industrial Data & Information Inc.

Glossary of Supply Chain Terminology

Word	Type	Definition of Word
Tracer	Tr	A form used to implement tracing and information gathering about a lost or misdirected shipment. (KA)
Traffic	Tr	1) Department/division responsible for obtaining most economic commodity classification and method of transportation materials and products. 2) People and/or property carried by transportation companies. (Cnf)
Traffic Management	Tr	The selection of transport modes and of specific carriers within the modes. (KA)
Trailer on FlatCar	Tr	See Piggyback. Shipments moving TOFC receive special rates from tariffs provided for that class of traffic. Abbreviated TOFC. (Cnf)
Tramp	Tr	Vessel that does not operate along definite route on fixed schedule, but calls at any port where cargo is available. (Cnf)
Transponder	Tr	Electronic tag attached to a truck or other vehicle. Information is electronically stored in the tag and when a roadside reader reads the tag, its information becomes available.
Transship	Tr	A term commonly used to denote transfer of goods from one means of transportation to another. The re-handling of goods en route. (Cnf)
Trip Charter	Tr	Hiring vessel to haul cargo for special voyage. (Cnf)
Triple	Tr	A combination of vehicles that has a tractor and three trailers. (Cnf)
Trip Lease	Tr	An arrangement in which a regulated carrier "leases" or hires an owner/operator to make a single run. (Cnf)

Suggest New Words http://www.IDII.com/addword.htm

© 2004 Industrial Data & Information Inc.

Glossary of Supply Chain Terminology

Word	Type	Definition of Word
Truckload	Tr	Abbreviated TL. 1. Quantity of freight that will fill a truck. 2. Quantity of freight weighing the maximum legal amount for a particular type of truck. 3. The quantity of freight necessary to qualify for a truckload rate. (Cnf)
TSA	Tr	Transportation Security Administration. Part of the US Government that is now under the DHS.
ULD	Tr	Unit Load Device. A container utilized for air freight.
Ullage	Tr	Empty space present when container is not full. (Cnf)
Ultra Vires	Tr	To act without apparent authority. (WA)
Undercharge	Tr	Where a carrier or its trustee in bankruptcy files a claim against the shipper for that shipper's practice of paying below the carrier's listed tariff. As a result of this practice, several carriers went out of business and brought undercharge claims in the approximated amount of $32 billion. (WA)
Uniform Straight Bill of Lading	Tr	The Uniform Straight Bill of Lading is the contract of carriage published in the National Motor Freight Classification (NMFC) that governs all motor carriers listed as participants in that tariff. (WA)
Valuation, Actual	Tr	The actual value of goods shown on a bill of lading by the shipper when the rate to be applied depends on the value of those goods. (Cnf)
VAT	Tr	Value Added Tax.

Suggest New Words   http://www.IDII.com/addword.htm

© 2004 Industrial Data & Information Inc.

Glossary of Supply Chain Terminology

Word	Type	Definition of Word
Venue	Tr	The rule of civil procedure that states the geographic location of where an action can be brought. Not to be confused with subject matter jurisdiction that deals with the question of which court is empowered to hear the action. (WA)
VIN	Tr	Vehicle Identification Number.
Voyage Charter	Tr	Engaging services of cargo ship for specified trip from one port to another at established tonnage rate. (Cnf)
Wastage	Tr	Loss of goods due to handling, decay, leakage, shrinkage, etc. (Cnf)
Waybill	Tr	A document containing the description of goods, which are part of a common carrier freight shipment. Shows origin, destination, consignee/consignor, and amount charged. Copies of this document travel with goods and are retained by originating/delivering agents. Used by carriers for internal records and control, especially during transit. It is not a transportation contract. (Cnf)
Weigh-In-Motion	Tr	Technology that weighs a vehicle while the vehicle is in motion.
Weigh Station	Tr	Permanent station equipped with scales at which motor vehicles transportation property on public highways are required to stop for checking of gross vehicle and/or axle weights. Many states also use portable scales to comply with their weight limits. Often combined with port of entry facilities. (Cnf)

Suggest New Words http://www.IDII.com/addword.htm

© 2004 Industrial Data & Information Inc.

Glossary of Supply Chain Terminology

Word	Type	Definition of Word
Weight	Tr	In shipping, weight is qualified further as gross (weight of goods and container), net (weight of goods themselves without any container), and legal (similar to net, determined in such manner as law of particular country/jurisdiction may direct). See Gross Weight, Net Weight. (Cnf)
Weight Bumping	Tr	When a household goods carrier falsifies the weight of the shipment by adding extra weight to the trailer before weighing it on a track scale. (WA)
Wharfage	Tr	A charge assessed against a shipping line for using a wharf, or against freight handlers moving over the pier or dock.(KA)
Yard	Tr	Unit of track systems within certain area used for storing cars, loading/unloading freight, and making up trains over which movements not authorized by timetable or train order may be made. Subject to prescribed signals/regulations. (Cnf)

Suggest New Words http://www.IDII.com/addword.htm

© 2004 Industrial Data & Information Inc.

Warehouse

Terms & Words
Utilized In The
Warehousing Industry

Related Sections: Conveyor, MSDS, and Pallet

Suggest New Words http://www.IDII.com/addword.htm

© 2004 Industrial Data & Information Inc.

Glossary of Supply Chain Terminology

Word	Type	Definition of Word
ABC	W	ABC classification. A user defined strategy for grouping parts within the warehouse (or organization). The grouping strategy is usually by product velocity, where fast movers are A items.
ABM	W	Activity Based Management
Affinity	W	Affinity code. A user defined code for products that look alike. E.G., 10" and 12" adjustable chrome wrenches.
AGV	W	Automated Guided Vehicle.
AG/VS	W	Automated Guided Vehicle System.
Aisle	W	A corridor or passageway within a warehouse, which allows access to, stored goods. (KA)
Aisle Space Percentage	W	Abbreviated ASP. The proportion of a warehouse devoted to aisles. (KA)
Allocated Stock	W	Inventory that has been reserved, but not yet picked from stock, and thus is not available for other purposes.
Allocation	W	Allocating inventory to a specific order. Order may be an outbound order (sales order, transfer order, work order, etc.). See Hard Allocation and Soft Allocation.
Alternate Commodity	W	A synonym for a Commodity, used by someone other than the owner of the product (i.e. the vendor's commodity name)
Anniversary Billing	W	A method of public warehouse billing for storage in which the customer is billed a one-month storage charge for all products as they are received. The same unit, if still in storage, is billed an additional monthly charge on each monthly anniversary date thereafter. This method does not involve any pro-rating of time in storage and so requires that anniversary dates for each item in storage be separately identified. See split-month billing. (KA)

Suggest New Words http://www.IDII.com/addword.htm

© 2004 Industrial Data & Information Inc.

Glossary of Supply Chain Terminology

Word	Type	Definition of Word
Annual Inventory	W	See Physical Inventory (KA)
ANOVA	W	Analysis of Variance
AOQ	W	Average Outgoing Quality
AOV	W	Average Order Value
Apron	W	The area directly outside the dock door upon which delivery vehicles can be parked or positioned for loading and unloading. This area is of the correct depth to allow trailer floors to line up with the warehouse floor. (KA)
ASP	W	Aisle Space Percentage.
ASRS	W	See Automatic Storage & Retrieval System.
Assembly Area	W	A warehouse location where materials, components, or finished products are collected and combined. (KA)
ATO	W	Assemble To Order. A technique to postpone inventory that can be utilized as components and/or finished goods. Purpose is to reduce the inventory required on-hand but still keep service rate high, by assembling a finished goods product at the last moment.
Automated Storage & Retrieval System	W	Material Handling Equipment - specifically an automated, mechanized system for moving merchandise into storage locations and retrieving it from storage locations.
Available Stock	W	Inventory that is available to service immediate demand.
Banding	W	Strapping to hold shipment together. Normally banding is plastic or metal strapping.
Batch Pick, Batch Picking	W	A method of picking a single SKU to be distributed across multiple orders.

Suggest New Words http://www.IDII.com/addword.htm

© 2004 Industrial Data & Information Inc.

Glossary of Supply Chain Terminology

Word	Type	Definition of Word
Bay	W	A designated area within a section of a storage area outlined by markings on columns, posts, or floor. (KA)
Bay Storage	W	The use of a large designated area for storing merchandise. (KA)
BBS	W	Business Balanced Scorecard
Beginning Inventory	W	The inventory count at the beginning of the current period. See Ending Inventory and Physical Inventory. (KA)
Bin	W	Four-sided superstructure to be mounted on a pallet base, with or without a cover; also known as a box or container bin pallet. (P1)
Bin	Pa W	Storage space for inventory at rest or being staged. A bin may be a physical bin or a virtual bin.
Block Pattern	W	A method of storing merchandise on a pallet in a pattern to allow a stable pallet load. (KA)
Bonded Warehouse	W	1. A warehouse approved by the Treasury Department and under bond/guarantee for observance of revenue laws. 2. Used for storing goods until duty is paid or goods are released in some other proper manner. (Cnf)
Bottleneck	B W	Congestion or significant slow down at an area due to an inefficient resource or process.
Brick Pattern	W	A method of storing merchandise on a pallet in a pattern to accommodate items of unequal width or length. See Pinwheel Pattern. (KA)
BTO	W	Built to Order

Suggest New Words   http://www.IDII.com/addword.htm

© 2004 Industrial Data & Information Inc.

Glossary of Supply Chain Terminology

Word	Type	Definition of Word
Buffer Stock	W	Safety Stock. The stock held to protect against the differences between forecast and actual consumption, and between expected and actual delivery times of procurement orders, to protect against stock outs during the replenishment cycle. In calculating safety stock, account is taken of such factors as service level, expected fluctuations of demand and likely variations in lead time. See lead time inventories.
Bulk Storage	W	1) Storage of merchandise and materials in large quantities, usually in original or shipping containers. 2) Storage of unpackaged commodities. (KA)
Bumpers	W	Pieces of rubber located at the floor level of a dock opening to cushion the building from truck trailer impacts. (KA)
Canopy	W	A covering over the area outside a dock door, used to prevent rain, ice, or snow from interfering with truck loading. (KA)
Cantilever Rack	W	A storage rack in which the shelves rest on arms that extend from center posts. (KA)
Carousel	W	Material Handling Equipment - specifically an automated delivery system to move product from storage to picker, without any travel on picker. Two types of carousels are on the market, horizontal carousels and vertical carousels. Carousels are most productive when pick lists are downloaded from the WMS into the carousel system and optimized for multiple orders being picked at one time.
Carton	W	A protective outer case that contains products. E.G., carton, tote, pallet.
CFR	W	Case Flow Rack.

Suggest New Words http://www.IDII.com/addword.htm

© 2004 Industrial Data & Information Inc.

Glossary of Supply Chain Terminology

Word	Type	Definition of Word
Co-Managed Inventory	W	A support arrangement similar to Vendor Managed Inventory but where replacement orders for the vendor-owned stock are agreed by the user prior to delivery replenishment cycle. In calculating safety stock, account is taken of such factors as service level, expected fluctuations of demand and likely variations in lead time.
Commodity	W	One of the major outputs of the manufacturing effort, also known as product codes. Within a warehouse, an item uniquely identified to a particular customer.
Consumable Stock	W	A classification of stock used to describe items that are totally consumed in use - e.g. gift wrap, packing materials, etc.
Container	Tr W	Anything in which articles are packed. A standardized box used to transport merchandise particularly in international commerce. Marine containers are typically 8'X8' with length of 10', 30', or 40'. These containers may be transloaded from rail cars or ships onto a truck frame and delivered to their final destination. (KA)
Contract Warehouse	W	See also Third Party Warehouse and Public Warehouse.
Cooler	W	A refrigerated space that holds material above freezing but usually below 50°. (KA)
CPG	W	Consumer Product Goods or Consumer Packaged Goods.
Cross-Docking	W	A method of fulfilling orders that the movement of goods directly from receiving dock to shipping dock, bypassing the putaway of inventory to stock.
Cross-training	W	Providing training or experience in several different warehousing tasks and functional specialties in order to provide backup workers. (KA)

Suggest New Words http://www.IDII.com/addword.htm

© 2004 Industrial Data & Information Inc.

Glossary of Supply Chain Terminology

Word	Type	Definition of Word
CTO	Mfg W	Configure To Order. See also ETO and MTS.
CTP	W	Capable To Promise. See ATP / Available To Promise.
Cube Loading	Tr W	The process of loading merchandise onto pallets or other unit-loading techniques to allow several unit loads to be stacked for transportation. (KA)
Cube Utilization	Tr W	The percentage of space occupied compared to the space available. (KA)
Cubiscan	W	A device that cubes and weighs products that are placed unto it. Cubiscan is a registered Trademark of Quantronix company. See http://www.cubiscan.com or main website http://www.quantronix.com.
Customer Pickup	W	1) Merchandise picked up by the customer at the warehouse. 2) The act of picking up merchandise at the dock. AKA Will-Call or Walk-In. (KA)

Suggest New Words http://www.IDII.com/addword.htm

© 2004 Industrial Data & Information Inc.

Glossary of Supply Chain Terminology

Word	Type	Definition of Word
Cycle Counting	W	Cycle counting is the physical counting of stock on a perpetual basis, rather than counting stock periodically. A cycle is the time required to count all items in the inventory at least once. The frequency of cycle counting can be varied to focus management attention on the more valuable or important items or to match work processes. Some of the systems used are: ABC system with the highest count frequency for items with the highest annual usage value, Reorder system when stocks are counted at the time of order, Receiver system with counting when goods are received, Receiver system with counting when goods are received reached to confirm that no stock is held, Transaction system where stocks are counted after a specified number of transactions.
Date Code	Mfg W	A label showing the date of production. In the food industry, it becomes an integral part of the lot number. (KA)
DC	W	See Distribution Center.
Dead Stock	W	A product that does not move from its storage location for a long time. (KA)
Deep-Lane Storage	W	Storage of merchandise greater than one unit deep on one or both sides of an aisle. See Double-Deep Lane Storage and Single Deep. (KA)
Deflection	W	The sag, bend, or deformation of a rack beam platform or container side due to the weight of a load. (KA)
Dekitting, Dekit	Mfg W	When a work order has been canceled, the process to return the components back to inventory is called dekitting.
DIR	W	Drive In Rack.

Suggest New Words   http://www.IDII.com/addword.htm

© 2004 Industrial Data & Information Inc.

Glossary of Supply Chain Terminology

Word	Type	Definition of Word
Distribution Center	W	A modern warehouse that processes inventory based upon the direction of the corporate systems. Abbreviated DC
Dockboard	W	A device for bridging the gap between the warehouse floor and a vehicle's load bed. (KA)
Dock Face	W	The outside wall of the dock door area. (KA)
Dock Fire Door	W	Generally used with enclosed apron areas, this is a safely feature to protect the interior dock area from fires that may occur on the apron or in the trailer itself. (KA)
Dock Leveler	W	A manually or hydraulically operated plate, located at the dock entrance, that can be raised and lowered approximately one foot to accommodate varying trailer floor heights. (KA)
Dock Light	W	A flood light positioned so it illuminates the interior of a trailer, which is not obstructing loading and unloading activities. (KA)
Dock Plate	W	A moveable metal ramp that allows access to a rail car or trailer. (KA)
Dock Receipt	W	A receipt issued for a shipment at a pier or dock. (KA)
Double-Deep Lane Storage	W	Rack storage of merchandise two loads deep on one or both sides of an aisle. See Deep-Lane Storage and Single Deep. (KA)
Draft Curtain	W	A fire curtain that hangs from the warehouse ceiling to reduce the spread of fire. (KA)
Drive-In Rack	W	Storage rack which provides side rails to allow high stacking in deep rows. Unlike drive-through rack, it provides access only from the aisle. (KA)

Suggest New Words http://www.IDII.com/addword.htm

© 2004 Industrial Data & Information Inc.

Glossary of Supply Chain Terminology

Word	Type	Definition of Word
Drive-Through Rack	W	Storage rack which provides side rails to allow high stacking of products in deep rows and access to the from either end of the row. (KA)
Driveway Installation	W	A ramp located on the outdoor apron of the dock, used to raise and lower a truck trailer, so that its floor becomes level with the dock floor. (KA)
DSD	B W	Direct Store Delivery. See also FRM.
DTS	W	Dock To Stock.
Dumbwaiter	W	A miniature freight elevator with a car moved by a hand-operated pulley. It is used for carrying small parcels. (KA)
Earmarked Material	W	On-hand material that is reserved an physically identified rather than simply allocated in inventory controls. (KA)
ECR	B W	Efficient Customer Response.
EHCR	B W	Efficient Healthcare Consumer Response. To realize supply chain cost savings through the adoption of EDI, bar-coding, and other strategies such as Activity Based Costing. See also ECR.
EI	W	See Ending Inventory
Emergency Order	W	As differentiated by an order type, an emergency order is processed in an expedited fashion.
Ending Inventory	W	Abbreviated EI. A statement of on-hand inventory levels at the end of a period. See Beginning Inventory and Physical Inventory. (KA)
ER	W	Expected Receipt (EM)
ETO	Mfg W	Engineer-To-Order. A product that is designed only when an order requires it. These are usually highly engineered goods and non-commodity items. See also CTO and MTS.
Ex-Dec, X-Dec	W	Export Declaration Documents.

Suggest New Words   http://www.IDII.com/addword.htm

© 2004 Industrial Data & Information Inc.

Glossary of Supply Chain Terminology

Word	Type	Definition of Word
Expiration Date	W	Date after which merchandise can no longer be shipped. In Expiration date controlled inventory, inventory must be discarded after this date.
Facility	Mfg W	The physical plant and storage equipment. Permanent storage bins in a warehouse may be considered part of the facility, whereas material handling equipment may not. (KA)
Family Grouping	W	A method of sorting products, with similar characteristics, into families to be stored in the same area. (KA)
FCS	W	First Customer Ship
FEFO	W	Inventory allocation method, first-expired, first-out.
F/G	Mfg W	Finished Goods
FGI	W	Finished Goods Inventory
FIFO	W	Inventory allocation method, first-in, first-out. This method allocates first items stored are the first items utilized.
FILO	W	Inventory allocation method, first-in, last-out.
Finished Product Inventory	Mfg W	Products available for shipment to customers. AKA Finished Goods. (KA)
Fire Wall	W	A wall made of fire-resistant material to prevent a fire from spreading. (KA)
First In, First Out	W	Inventory allocation method, first-in, first-out. This method allocates first items stored are the first items utilized. Abbreviated FIFO
FISH	W	First In Still Here. A funny, but real expression of very slow movers that needs to be dealt with.
Flow Rack	W	In one side and out the other. Product stored in this way is necessarily FIFO within the rack itself.

Suggest New Words   http://www.IDII.com/addword.htm

© 2004 Industrial Data & Information Inc.

Glossary of Supply Chain Terminology

Word	Type	Definition of Word
Flow Through	W	See Cross Dock.
Flow Through Rack	W	In one side and out the other. Product stored in this way is necessarily FIFO within the rack itself.
FLT	W	Fork Lift Truck. See Forklift. (EM)
Forklift	W	A vehicle with horizontal lift that moves product & freight within a warehouse or on the dock. AKA Fork-Lift, Hi-Lo or Lift-Truck.
Forks	W	The flat metal appendages mounted on lift trucks to facilitate the movement of merchandise on pallets. They are generally four to six inches wide, and 42-48 inches long. (KA)
Forward Location	W	See Pick Slot.
FPO	W	Firm Planned Orders
Free-Standing Rack Structure	W	A storage rack supported only by the floor. Free standing racks are not attached to the ceiling, walls, or any other part of the warehouse. (KA)
Freezer	W	Temperature-controlled (below 32°) storage space for perishable items. Also Cooler. (KA)
FRM	W	Floor Ready Merchandise. Prepared products are received as "floor ready" at the retail store. Before shipping these goods, the warehouse or supplier will add prices, price stickers, tags, security devices, special packaging, etc. so that goods may rapidly be cross docked through RDCs or received directly at the retail stores.
FSL	W	Forward Stock Location. Primary location for picking of a product. AKA as a "pick face".

Suggest New Words http://www.IDII.com/addword.htm

© 2004 Industrial Data & Information Inc.

Glossary of Supply Chain Terminology

Word	Type	Definition of Word
Gantry Crane	W	Used in warehouse facilities where oversized or very heavy items such as pipe, steel, and heavy machinery must be loaded onto trucks or rail cars. (KA)
Gaylord	W	Large corrugated carton that is sized to fit a pallet's length and width. Typically used for loose parts or other items stored in bulk. Most commonly found in manufacturing.
Girth	W	The measurement around the sides of the container. (KA)
Godown	W	A waterfront storehouse in the Orient. (KA)
GOH	W	Goods on hanger. Some wearing apparel is shipped on hangers without packaging. (KA)
Gross Weight	W	Weight of inventory item with packing and container. See also Gross Weight.
Handling	W	The movement of materials or merchandise within a warehouse. (KA)
Handling Charge	W	A charge for ordinary warehouse handling operations. (KA)
Handling Costs	W	The costs involved in warehouse handling. (KA)
Hand Truck	W	A device used for manually transporting goods. A metal plate is slid under the load, then truck and load are tilted toward the operator and moved. There are two varieties: The western type has its wheels located within the side rails, while the eastern type places the wheels located outside the side rails. (KA)

Suggest New Words http://www.IDII.com/addword.htm

© 2004 Industrial Data & Information Inc.

Glossary of Supply Chain Terminology

Word	Type	Definition of Word
Hard Allocation	W	Detailed allocation of inventory that specifies the exact bin, license plate, lot, and/or serial numbers. Hard allocations are done by the warehouse management system at wave release or pick release. Soft allocation has been done previously, which committed a specified quantity only.
HazMat	Tr W	Hazardous Material. Product is classified hazardous by a government agency or a carrier. Special handling is required on some hazardous items, which may include segregation, separation wall, temperature control, and/or limitations on how the product may be shipped. See also MSDS and Hazard Class.
High	W	Number of Layers per Pallet
Hi-Lo	W	Another name for a forklift, or a vehicle with horizontal lift. A vehicle to move freight within a warehouse or on the dock. AKA Fork-Lift, Towmotor, or Lift-Truck. See Fork-Lift.
HMI	Mfg W	Human-Machine Interface
Hoist	W	An apparatus for lifting or lowering a load. (KA)
Honeycomb Factor	W	A term that describes the amount of storage space lost to partially-filled rows or to pallet positions not fully filled. (KA)
Honeycombing	W	A waste of space that results from partial depletion of a lot and the inability to use the remaining space in this area. (KA)
Horizontal Carousel	W	See Carousel.
Housekeeping	W	Maintaining an orderly environment for preventing errors and contamination in the warehousing process. Good housekeeping includes keeping aisles clear, trash disposal, sweeping, orderly stacking, and removal of damaged goods. (KA)

Suggest New Words http://www.IDII.com/addword.htm

© 2004 Industrial Data & Information Inc.

Glossary of Supply Chain Terminology

Word	Type	Definition of Word
HU	W	Handling Unit.
IDEAS	W	Internet-based Data Envelopment Analysis System. An online benchmarking tool for warehouse and distribution center to get a "system view" of their warehouse performance. This is a self-assessment tool – developed in 2001 by Georgia Tech and MHIA. IDEAS are located at http://www.isye.gatech.edu/ideas.
IFO	W	Income From Operations
Inner Packaging	W	Materials such as paper, foam, or wood shavings used to cushion impacts and prevent the movement of goods within a container. (KA)
Interleave	W	Multi-tasking. The warehouse worker will be directed to do multiple types of tasks in one trip.
Internal Costs	W	Those costs generated within the facility and directly under the control of warehouse management these include storage, handling, clerical services, and administration. (KA)
KDF Cartons	W	Knocked down (flat) carton – unassembled packages. (KA)
Kit	W	A number of separate Stock Keeping Units that are supplied or used as one item under its own Item Number.
Kitting	W	Light assembly of component parts, often performed in a warehouse. (KA)
Knock-Down	W	Abbreviated KD. When articles are taken apart for the purpose of reducing the cubic space of the shipment, the disassembly process is referred to as a knock-down shipment. (KA)

Suggest New Words http://www.IDII.com/addword.htm

© 2004 Industrial Data & Information Inc.

Glossary of Supply Chain Terminology

Word	Type	Definition of Word
Last In, First Out	W	Inventory allocation method called Last In, First Out. Abbreviated LIFO
Layer	W	A group of Commodities residing in one horizontal dimension on a Pallet. Each Pallet consists of one or more Layers. During picking, Commodities are generally depleted from one Layer completely, before the next layer is taken.
Layout	W	The design of the storage areas and aisles of a warehouse. (KA)
LES	Tr W	Logistics Execution Systems. Involves transportation management (TMS) and warehouse management (WMS) software. AKA Supply Chain Execution (SCE) or Logistics Resource Management (LRM).
Let Down	W	Replenishment. Moving inventory from reserve storage to the active picking slots below. Normally the reserve storage is above the forward picking slot, therefore this replenishment is called a "let down".
LF	W	Logistics Fulfillment (EM)
License Plate Number	W	A unique number that is applied to a container to rapidly identify the contents (products) within. The LPN is normally printed as a barcode on a label or the unique id of the RFID tag.
LIFO	W	Inventory allocation method called Last In, First Out.
Lift Truck	W	Another name for a forklift, or a vehicle with horizontal lift. A vehicle to move freight within a warehouse and dock. AKA Fork-Lift, Towmotor, or Hi-Lo. See Fork-Lift.
Lineside Warehouse	W	A supplier warehouse positioned as close as possible to the production location to facilitate Just In Time manufacture.

Suggest New Words   http://www.IDII.com/addword.htm

© 2004 Industrial Data & Information Inc.

Glossary of Supply Chain Terminology

Word	Type	Definition of Word
Live Rack	W	A storage rack constructed to allow items to move unaided toward the picking point. The rack is slanted so that the picking point is lower than the rear loading point, allowing gravity to draw items to the front. A roller conveyor or other low-friction surface supports the merchandise. (KA)
Location	W	The place in a warehouse where a particular product can be found. (KA)
Location Audit	W	A systematic verification of the location records of an item or group of items by checking the actual locations in a warehouse or storage area. (KA)
Logistics	Tr W	Logistics plans, implements, and controls the efficient, effective forward and reverse flow and storage of goods, services, and related information between the point of origin and the point of consumption in order to meet customers' requirements. (Cnf)
Loose	W	The number of remaining Commodities left on a particular Layer. Ex. Tie, High Loose: 7, 3, 1 means 7 boxes per layer, 3 layers high, and 1 loose box on top of the 3rd layer.
Lot	W	Product that has a life span with an expiration date, Manufacturing date, and/or code date. A lot is a group of an item that has been made using the same ingredients (like a batch), a production run, or some other grouping to be used for identification purposes, unique from other same items produced in a different batch or run. For instance fabric color in the same SKU may not match, lot to lot.
Lot Number	W	Identifying numbers used to keep a separate accounting for a specific lot of merchandise. (KA)

Suggest New Words http://www.IDII.com/addword.htm

© 2004 Industrial Data & Information Inc.

Glossary of Supply Chain Terminology

Word	Type	Definition of Word
Lot Number Traceability	W	The ability to track an item or group of items by a unique set of numbers based by a vendor or production run. (KA)
Lottable	W	Specific attributes of a Commodity that, taken together, differentiates like Commodities, and allow like Commodity to be tracked as separate Lots throughout the facility. Owners of the Commodity drive what is considered a Lottable. Examples of Characteristics: color, pack size, original production facility, etc.
LPN	W	License Plate Number. A unique number assigned to a container. A container is any type of container that can hold inventory – such as a pallet, carton, and tote. WMS use & assign LPNs to track inventory.
LRM	Tr W	Logistics Resource Management. Software that includes modules with transportation management, warehouse/inventory management, and contracting/rate negotiations. AKA as Logistics Execution System (LES) and Supply Chain Execution (SCE).
Marking Machine	W	A machine that imprints or embosses a mark on a label, ticket, tape, package, or tag. (KA)
Master Carton	W	A single large carton that is used as a uniform shipping carton for smaller packages. It is used primarily for protective purposes, but also simplifies materials handling by reducing the number of pieces handles. See Master Pack. (KA)
Master Pack	W	An established quantity for a manufacturer's product. The master pack is a carton containing a set number of multiple case quantities.

Suggest New Words http://www.IDII.com/addword.htm

© 2004 Industrial Data & Information Inc.

Glossary of Supply Chain Terminology

Word	Type	Definition of Word
Material Handling	W	The movement of materials going to, through, and from warehousing, storage, service facility, and shipping areas. Materials can be finished goods, semi-finished goods, components, scrap, WIP, or raw stock for Manufacturing.
MHE	W	Material Handling Equipment.
MHE	W	Mechanical Handling Equipment.
Master Pack	W	An established quantity for a manufacturer's product. The master pack is a carton containing a set number of multiple case quantities.
Mat	Pa W	A panel of wood, rubber, or other material that is placed on top of unit loads to allow tight strapping of the load without product damage. (KA)
Materials Management	W	The functions that define the complete cycle of material flow, from the purchase to the distribution of the finished product. This includes functions of planning production materials, control of work in process, warehousing, shipping, value added services at the warehouse, and distribution.

Suggest New Words http://www.IDII.com/addword.htm

© 2004 Industrial Data & Information Inc.

Glossary of Supply Chain Terminology

Word	Type	Definition of Word
Material Safety Data Sheet	W MSDS	A Material Safety Data Sheet (MSDS) is prepared by a chemical manufacturer, and summarizes available information on the health, safety, fire, and environmental hazards of a chemical product. It also gives advice on the safe use, storage, transportation, and disposal of that product. Other useful information such as physical properties, government regulations affecting the product, and emergency telephone numbers are provided in the MSDS as well. There is a detailed description of how to read an MSDS and a useful glossary of MSDS terms in Hach's Website at http://www.hach.com. (HC)
Materials Warehouse	W	A warehouse used exclusively for the storage of raw materials. (KA)
MCS	W	Material Control Software. AKA Material Handling Control Software, Warehouse Control Systems, or Conveyor Control Software. This is the software that interfaces the machinery to the WMS or MES.
Mesh	W	A panel or fabric with openings (pores) of uniform size. It is designed by citing the number of openings per square inch. (KA)
Min-Max System	W	An order-point replenishment system. The minimum point is the order point and the maximum is the "order-up-to" level. The order quantity is variable to take advantage of usage patterns and lot-sized ordering economies. (KA)
MLP	W	Master License Plate (EM)
MRO	W	Maintenance, Repair, and Operations.
MSDS	W MSDS	See Material Safety Data Sheet.

Suggest New Words   http://www.IDII.com/addword.htm

© 2004 Industrial Data & Information Inc.

Glossary of Supply Chain Terminology

Word	Type	Definition of Word
MTO	Mfg W	Make To Order. A product that is manufactured only when an order is confirmed for it. These are usually low volume or highly engineered goods. See also CTO, ETO, and MTS.
MTS	Mfg W	Make To Stock. Products are manufactured and placed into warehouse as finished goods. The quantity built is based upon safety stock and re-order point calculations. See also CTO, ETO, and MTO.
Mullen Test	W	A device to test, or the process of testing packaging material to establish its strength. (KA)
Multi-Tine Fork	W	Attachment to a forklift truck that allows the movement of two pallets side-by-side, rather than one pallet at a time. (KA)
Negotiable Warehouse Receipt	W	A legal certification that listed goods are held in public warehouse. The certificate can be purchased or sold, thus transferring title to the goods. See Non-negoiable Warehouse Receipt. (KA)
Nested	W	The process of packing articles so that one rests partially or entirely within another, thereby reducing the total cubic displacement. (KA)
Nested Solid	W	Articles nested so that the bottom of one rests on the bottom of the lower one. (KA)
Nesting Factor	W	Nesting on products that "fit" within each other, therefore the first item is full cube, the 2nd, 3rd just use the "factor" cube. E.G., Buckets would have a .20 (20%) nesting factor.

Suggest New Words http://www.IDII.com/addword.htm

© 2004 Industrial Data & Information Inc.

Glossary of Supply Chain Terminology

Word	Type	Definition of Word
Net Storage Area	W	The amount of warehouse space actually used for storage of merchandise. It usually expressed in square feet or meters and excludes aisles, dock areas, staging areas, and offices. (KA)
Net Weight	W	Weight of inventory item without packing and container. See also Gross Weight.
NIS	W	Not In Stock.
Non-Negotiable Warehouse Receipt	W	A legal certification that the goods listed on it are now in the custody of the public warehouse. The certificate cannot be bought or sold. (KA)
Non-Read	C W	Failure of scanner to read a barcode or RFID tag.
NSN	W	National Stock Number. See also SKU.
Obsolete Stock	W	Stock held within an organization where there is no longer any reason for holding the stock.
On-Hand Balance	W	The recorded level of inventory in a distribution center. (KA)
Order Lead Time, Order Cycle Time	W	The total internal processing time necessary to transform a replenishment quantity into an order and for the transmission of that order to the recipient.
Order Picker	W	1) Lift truck which allows the warehouse worker to ride with the pallet and to pick from various levels. 2) A warehouse worker whose prime job is selection of orders. (KA)
Order Picking	W	Collecting items from a storage location or picking location to satisfy an order.
OSD	Tr W	See Over, Short, and Damaged.
Over, Short, and Damaged	Tr W	A report issued at the warehouse when goods are damaged. Used to file a claim with the carrier. (Cnf)

Suggest New Words   http://www.IDII.com/addword.htm

© 2004 Industrial Data & Information Inc.

Glossary of Supply Chain Terminology

Word	Type	Definition of Word
Overage	Tr W	Freight in excess over quantity believed to have been shipped, or more than quantity shown on shipping document. (Cnf)
Overhead Cost	W	Those costs that are not directly related to warehousing and storage, which are still part of the total costs of a facility. These include janitorial services, heat, light, power, maintenance, depreciation, taxes, and insurance. (KA)
Overshipment	Pu W	A shipment containing more than originally ordered. (KA)
Pack Size	W	A collection of Units per Measure that describe how a Commodity is packaged. For example, Commodity ABC comes packaged in 24 each per Case, and 50 Cases per Pallet.
Packing and Marking	W	The process of packaging goods for safe shipment and handling, and appropriately labeling the contents. (KA)
Packing List	W	List showing inventory that was packed in shipment. Prepared by warehouse personnel. Copy is sent to customer / consignee to help verify shipment received. Packing list detail can be also transmitted to customer / consignee via EDI Advance Ship Notice, if containers are uniquely coded.
Palletization	Pa W	System for shipping goods on pallets. Permits shipment of multiple units as one large unit. (Cnf)
Palletize	Pa W	To place material on a pallet in a prescribed area. (KA)
Palletizer	Pa W	A type of materials-handling device suing conveyor or robotics to position cubes or bags on a pallet. (KA)
PBR	W	Push back rack.

Suggest New Words http://www.IDII.com/addword.htm

© 2004 Industrial Data & Information Inc.

Glossary of Supply Chain Terminology

Word	Type	Definition of Word
PDM	W	Product Data Management. Software module that is focused on proper setup, maintenance, and changes to the life of the product.
PDT	C W	Portable data terminals. A rugged hand-held computer with the computing power of a stationary computer. Normally equipped with RF (radio frequency) and an antenna.
Perpetual Inventory System	W	An inventory control system where a running record is kept of the amount of stock held for each item. Whenever an order is filled, the withdrawal is logged and the result compared with the re-order point for any necessary re-order action.
PF&D	W	Personal fatigue, and delay times. This is expressed as a percentage of time allowed for completing a task. (KA)
PFR	W	Pallet Flow Rack.
Physical Inventory	W	A physical count of every item located within the warehouse. AKA Annual Inventory. Cycle counts can replace physical counts. (KA)
PI	W	Physical Inventory
Pick-and-Pass	W	Picking technique used with pick zones, flow racks & conveyors. One picker will pull products into tote (or other container). Tote will be passed to next zone for picker to pull and put products into same tote. This pick-and-pass continues until all zones are completed and then tote is taken away on a take-away conveyor.
Picker	W	Term used for the person assigned the job of locating and removing stock from storage locations. See Order Picker. (KA)

Suggest New Words   http://www.IDII.com/addword.htm

© 2004 Industrial Data & Information Inc.

Glossary of Supply Chain Terminology

Word	Type	Definition of Word
Pick List	W	An output from a warehouse management system designating those items, by item number, description and quantity, to be picked from stock to satisfy customer demand.
Pick Time	W	The amount of time it takes a worker to select and document an item. (KA)
Pick-to-Carton	W	Picking technique where picker is directed to put pulled product directly into a licensed plated container (a "carton" normally). Picker may be picking for multiple orders during one trip, by using multiple cartons. When finished the picker puts the cartons on a take away conveyor for the shipping station.
Pick-to-Light	W	Picking technique used where pickers are directed by lights and/or digital displays at each bin. Very effective technique for high volume piece picking. Requires the WMS to be interfaced to Pick-To-Light MHE interface.
Pick Slot	W	The location where inventory is stored that is dedicated or picking. The pick slot is commonly replenished via a minimum trigger point to initiate replenishment. The terms forward location, primary location, and picking face are synonyms for this term.
Picking Face	W	See Pick Slot.
Pinwheel Pattern	W	A method of storing merchandise on a pallet in a pattern to arrange items of unequal width or length. See Brick Pattern and Row Pattern. (KA)
Portable Plate	W	A loading ramp that can be removed to any loading position on the dock. (KA)

Suggest New Words http://www.IDII.com/addword.htm

© 2004 Industrial Data & Information Inc.

Glossary of Supply Chain Terminology

Word	Type	Definition of Word
Post Audit	W	A study conducted of a new warehouse, fleet, or equipment to ascertain how well it is performing in relation to the proposal and financial analyses used to originally justify it. (KA)
Postponement of Final Assembly	W	The delay of final assembly until a firm customer order is received. Often, common parts and components can be produced, shipped, and inventoried at lower cost and risk than that associated with completed products. (KA)
Pressure Label	W	A price ticket or other information ticket that can be affixed to merchandise by pressing it on. (KA)
Primary Location	W	See Pick Slot.
Private Warehouse	W	A warehouse operated by the owner of the goods stored there. A private warehouse can be an owned or leased facility. (KA)
Pro Number	Tr W	A unique number assigned to a freight shipment by the carrier. This Pro Number is utilized on the freight bill and bill of ladings.
PTL	W	Pick To Light technology uses displays at each location that light up to show where to pick. Picker goes to this location, selects quantity displayed and pushes a button to confirm the pick. Used in high volume items. Cost is estimated to be $200 per light position in 2001.
Public Warehouse	W	See also Contract Warehouse and Third Party Warehouse.

Suggest New Words http://www.IDII.com/addword.htm

© 2004 Industrial Data & Information Inc.

Glossary of Supply Chain Terminology

Word	Type	Definition of Word
Push-Back Rack	W	Rack system that allows palletized product to be stored by being pushed up an inclined ramp. This allows for deep pallet storage.
Putaway	W	The movement of material from the point of receipt to a storage area. (KA)
QA	W	See Quality Assurance/Quality Control.
QC	W	See Quality Assurance/Quality Control.
QR	W	Quality Review. Check procedure on randomly selected components and/or finished goods product. See also QA, QC.
QS	W	Quality Standard. See Automotive Industry Action Group requirements. (EM)
Quality Assurance, Quality Control	W	Process of inspecting merchandises being received or shipped to ensure that the merchandise is of adequate quality and that the case content specification is accurate.
Quarantine	W	An inventory status indicating that this product is not available for inventory allocation but is in process of inspection for QA or another reason. See also QA and QC.
Quarantine	W	The isolation of goods or materials until they can be checked for quality or conformance with all required standards. (KA)
Rack	W	A structured storage system (single-level or multi-level) that is used to support high stacking of single items or palletized loads. (KA)
Rack-Supported Building	W	A warehouse in which the storage rack functions as the structural support for the roof. (KA)

Suggest New Words http://www.IDII.com/addword.htm

© 2004 Industrial Data & Information Inc.

Glossary of Supply Chain Terminology

Word	Type	Definition of Word
Radio Frequency	W	A system of mobile devices used to issue tasks, edit entered information and confirm the completion of tasks using either laser scanners or terminal entry. RF allows operations personnel to function in a "point-of-work" mode.
Ramp	W	An inclined roadway that connects levels in the warehouse. (KA)
Random Storage	W	One or more areas of the warehouse that are designed for random storage. Bins in a random storage area are not pre-assigned, but the WMS is allowed to fully evaluate what inventory to locate in these random storage bins.
RDC	W	Retail Distribution Center. A DC replenishing products directly to the stores. See Distribution Center.
Reach Truck	W	A forklift with the extended ability to reach forward significantly. Very useful for getting or retrieving the deeper pallet in a double deep pallet bin.
Real Time	W	Real time means that the new or updated information is instantaneously saved in the database.
Reasonable Care	W	The extent to which a warehouse operator is liable for goods. As defined in section 7-240-1 of the Uniform Commercial Code: A warehouseman is liable for damages or injury to the goods caused by his failure to exercise such car in regard to them as a reasonably careful person would exercise under like circumstances, but unless otherwise agreed he is not liable for damages that could not have been avoided by the exercise of such care. (KA)
Receiving Report	W	A record of the condition in which merchandise arrived. (KA)

Suggest New Words   http://www.IDII.com/addword.htm

© 2004 Industrial Data & Information Inc.

Glossary of Supply Chain Terminology

Word	Type	Definition of Word
Receiving Tally	W	The warehouse receiver's independent listing of goods unloaded from an inbound vehicle, sometimes prepared on a blind basis to ensure accuracy. (KA)
Refrigerated Warehouse	W	A warehouse that provides refrigeration and temperature control for perishable products. (KA)
Release	W	The authorization to ship material. (KA)
Renewal Storage	W	The rebilling fee (usually monthly) for products stored in a public warehouse. (KA)
Repack	W	Task that takes product in one configuration and is repackaged into another. 8 packs into 12 packs.
Replenishment	W	The task initiated to fill a picking location. Typically a replenishment task can be specific to an item's replenishment quantity and unit of measure.
Reverse Logistics	B W	The requirement to plan the flow of surplus or unwanted material or equipment back through the supply chain after meeting customer demand.
Rewarehousing	W	The process of calculating the best slotting positions for inventory and moving inventory to those optimized bins. This is done to optimize space utilization and decrease deadheading travel time.
RF	W	See Radio Frequency.
RFDC	W	Radio Frequency Data Collection (EM)
RFID	W	Radio Frequency Identification. Usually refers to RFID tags and readers.
RPC	W	Reusable Plastic Container.
RTLS	Tr W	Real Time Locating System. Usually in reference to a WMS or YMS ability to move-track-locate inventory.

Suggest New Words http://www.IDII.com/addword.htm

© 2004 Industrial Data & Information Inc.

Glossary of Supply Chain Terminology

Word	Type	Definition of Word
SCC, SCC-14	Std W	Shipping Container Code consisting of 14 digits. This UCC SCC-14 code is frequently put on a pallet or container label. The first two digits are the UCC application identifier. The next digit is the packaging level indicator. The next seven digits is the manufacturer identification number. The next five digits represent the manufacturer assigned product number. The last digit is a modulus 10 check digit. See also UCC 128.
SCE	W	Supply Chain Execution. Refers to the "execution" side of fulfilling the customer order and supporting functions to do so. Commonly includes warehouse, transportation, data interchange.
SCES	C W	Supply Chain Execution Systems. Software and processes to enable then execution side of fulfilling customer orders and supporting functions. Includes warehouse operations, transportation, with heavy emphasis on inventory & order entry.
SCU	W	Speech Control Unit. The Speech Control Unit is a computer that connects to the WMS for exchange of information. The SCU then gives individual instructions to the SDT via a wireless network. See also SDT.
SDT	W	Speech Data Terminal. In voice-activated devices, mobile workers use small, unobtrusive wearable computers called Speech Data Terminals with an attached headset. This would be used in a pick-to-voice environment. See SCU.
Seasonal Inventory	Pu W	Inventory held to meet seasonal demand. (KA)

Suggest New Words http://www.IDII.com/addword.htm

Glossary of Supply Chain Terminology

Word	Type	Definition of Word
Semi-Finished Inventory	Mfg W	Materials that are no longer in raw-material form, but which have not completed the production cycle to become finished goods. (KA)
Sequencing	W	The process of organizing items in a load so they will be in the order needed for production. (KA)
Serial Number	W	A unique identification number assigned to a single item. (KA)
Serpentine	W	A picking path that is in a serpentine pattern of bins to inventory from. Some call this a Z picking path.
Shelf Life	W	The length of time a product can be kept for sale or use before quality considerations make it necessary or desirable to remove it. (KA)
Shelter	W	A cover that protects the space between the door of a rail car or truck and a warehouse from inclement weather. (KA)
Ship-Age Limit	W	The final date a perishable product can be shipped to a customer. (KA)
Shipment	Tr W	1. Lot of freight tendered to carrier by one consignee at one place at one time for delivery to one consignee at one place on one bill of lading. 2. Goods/merchandise in one or more containers, pieces, or parcels for transportation from one shipper to single destination. (Cnf)
ShipTo	W	The name and address of where the shipment will be delivered. Also known as the consignee.
Shop Floor Control	Mfg W	The process of monitoring and controlling production or warehousing activities to ensure that procedures are followed. (KA)
Shrinkage	W	Reduction in bulk measurement of inventory. (KA)

Suggest New Words http://www.IDII.com/addword.htm

© 2004 Industrial Data & Information Inc.

Glossary of Supply Chain Terminology

Word	Type	Definition of Word
Shrink Wrap	Tr W	A plastic wrap used by shippers to secure cartons on a pallet. (Cnf)
Shroud	W	A protective sheet that covers the top and sides of a load, but which permits air to circulate from the bottom. (KA)
Single Deep	W	A bin. See also Double Deep.
Site Selection Model	W	A program that helps to determine the best location for a distribution center, or other facility. (KA)
SKU	W	Stock Keeping Unit. A product or a set of products referenced by the manufacturer by a unique part number.
Slip Sheet	W	Sheet of plastic, cardboard, or fiberboard used rather than a pallet. Forklift attachment is required to handle pallets with slip-sheets.
Slot	W	A position within a storage area reserved for a particular SKU. See also Bin. (KA)
Soft Allocation	W	Initial allocation of inventory to a specific line item of an order. A "soft" allocation just commits this inventory to the order, but does NOT specifically identify the bin, lot, serial number, or license plate. Soft allocation is done by the ERP system normally. See Hard Allocation.
Sortation	W	The process of separating packages according to their destination. (KA)
Split-Month Billing	W	A method of public warehouse billing for storage in which the customer is billed for all inventory in the warehouse at the beginning of the month, as well as for each unit received during that month. Merchandise received during the first half of the month is billed at a full-month storage rate, while merchandise received after the 15th day of the month is billed at a half-month storage rate. See Anniversary Billing. (KA)

Suggest New Words http://www.IDII.com/addword.htm

© 2004 Industrial Data & Information Inc.

Glossary of Supply Chain Terminology

Word	Type	Definition of Word
Split Shipment	W	A partial shipment that occurs when a warehouse is unable to fill an entire order. The remainder of the order is backordered. (KA)
Spoilage	W	1) One form of product deterioration. 2) The reduction in an inventory's value resulting from inadequate preservation or excess age. (KA)
Spot Check	W	A method of inspecting a shipment in which only a sampling of the total number of containers or items are received are inspected. (KA)
Stacked Loads	W	Unit loads on pallets that are placed on top of each other to created a column of unitized loads. (KA)
Stacking	W	The process of placing merchandise on top of other merchandise. (KA)
Stacking Height	W	The distance as measured from the floor to a point 24 inches or more below the lowest overhead obstruction. Stacking height is usually controlled to maintain clearances required by fire regulations. (KA)
Stacks	W	Refers to product stacked in the warehouse. (KA)
Staging Area	W	Temporary storage in a warehouse or terminal where goods are accumulated adjacent to the dock for final loading. (KA)
Stock	W	1) Term used to refer to inventory on hand. 2) The activity of replenishing merchandise in storage. (KA)
Stock Keeping Unit	W	Abbreviated SKU. A product or a set of products referenced by the manufacturer by a unique part number.
Stock Locator System	W	A system that allows all storage spots within a warehouse to be identified with an alpha-numeric code, and tracks the items and quantity in each location. (KA)

Suggest New Words http://www.IDII.com/addword.htm

© 2004 Industrial Data & Information Inc.

Glossary of Supply Chain Terminology

Word	Type	Definition of Word
Stockout	W	An event that occurs when one is out of stock on a specific product. Some software solutions record this event for information and/or input to purchasing.
Stock Picker	W	See Order Picker. (KA)
Stock Rotation	W	The process of moving or replacing merchandise to insure freshness and to maximize shelf life. (KA)
Storage Charge	W	A fee for holding goods at rest. (KA)
Storage Costs	W	The sum total of all costs associated with storage, including inventory costs, warehouse costs, administrative costs, deterioration costs, insurance, and taxes. (KA)
Storage Rate	W	The price charged for storage of merchandise, expressed as a cost per unit per month, or as a cost-per-square-foot (or meter) per month. (KA)
Strap Loading	W	The process of loading merchandise onto a pallet and securing it with metal or plastic straps. (KA)
Strapping	W	A metal or plastic band used to hold cases together in a unit. (KA)
Stretch Wrap	W	A process and means of applying a sheet of flexible plastic to packages in such a way that they are secured together in a unitized load. (KA)
Tag	W	A method of identifying an item or shipment. (KA)
Tally	W	A sheet made up when goods are received to count and record their condition on arrival. (KA)
Tare	W	The weight of packaging or containers. Tare weight plus net weight equals gross weight. (KA)

Suggest New Words   http://www.IDII.com/addword.htm

Glossary of Supply Chain Terminology

Word	Type	Definition of Word
Terminals	W	The term for warehouses in early transportation systems. These storage facilities were at the terminal points for land and sea sport. (KA)
Third Party Warehouse	W	A warehouse operated by a 3PL that contains the client's inventory. See also 3PL, Public Warehouse, and Contract Warehouse.
Throughput	W	1) The total number of units arriving at and departing from warehouse divided by two. Used in public warehouse rate making to calculate average movement of product. 2) A measure of the amount of work done by a computer. Throughput is dependent on hardware and software, and is more useful than simple measure of hardware speed. (KA)
Tie	W	Number of Units per Layer
Tiedown	W	A system of securing a unit load to a pallet. (KA)
Tier	W	1) A single layer of boxes or bags forming one layer of a unitized load. 2) A set of storage locations that are the same height. (KA)
Tine	W	The horizontal load-lifting portion of a fork on a fork truck. It is the portion of a fork that contacts the load. (KA)
TO	B W	Transfer Order. An internal company order to move inventory from one part of the company to another. This could be stock transfer between warehouses, or warehouse to a van, or division to division.
TOR	W	Top of Roller.
TowMotor	W	Another name for a forklift, or a vehicle with horizontal lift. A vehicle to move freight within a warehouse or on the dock. AKA Fork-Lift, Hi-Lo or Lift-Truck. See Fork-Lift.

Suggest New Words http://www.IDII.com/addword.htm

© 2004 Industrial Data & Information Inc.

Glossary of Supply Chain Terminology

Word	Type	Definition of Word
Traceability	W	The ability to track a shipment or an item. Any item with a lot number or serial number should be traceable back to the manufacturer, date and location of assembly. See also Lot Number Traceability. (KA)
Trackable	W	The collection of Quantity, Gross Weight, Net Weight, Space, and Pallet.
Truck Door	W	The part of the warehouse, which accommodates loading and unloading of trucks. It includes an overhead door, and may include a dock leveler, a dock shelter, and a concrete pad for the trailer. AKA Dock Door. (KA)
Turrett Truck	W	Material handling equipment on which the fork rotate 180 degrees to store or retrieve pallets from either side of the vehicle in a narrow aisle. (KA)
ULD	W	Unitized Loading Device. Freight is shipped in a ULD for loading in and out of aircraft.
Unitization	W	1) The consolidation of a number of individual items onto one shipping unit for easier handling. 2) The securing or loading of one or more large items of cargo into a single structure or carton. (KA)
Unitize	W	To consolidate packages into a single unit by banding, binding, or wrapping. (KA)
Unit of Measure	W	The degree of detail to which we refer about a Commodity. These are usually expressed as Eaches, Cases, Innerpacks, and Pallets. Also known as UOM.

Suggest New Words http://www.IDII.com/addword.htm

© 2004 Industrial Data & Information Inc.

Glossary of Supply Chain Terminology

Word	Type	Definition of Word
Unit per Measure	W	A count expressed as a smaller Unit of Measure per so many larger Units of Measure. For example, There are 24 Cans (Eaches) per 1 Case. Cans are the smaller UOM, since more cans fit into the larger Case.
UOM	W	See Unit of Measure.
Uprights	W	Vertical numbers used in storage rack. (KA)
Vacuum Packaging	W	The process of sealing packages by removing nearly all air. (KA)
VAL	W	Value Added Logistics. See Value Added Services.
Value Added	W	The contribution made by a step in the distribution process to the functionality, usefulness, or value of a product. (KA)
Value Added Services	W	Often the distribution center is required to perform services for customer orders other than picking and packing. These services are termed Value Added Services and they include (but are not limited to): Ticketing, Kit Assembly, Packaging, Final Finishing, Private Labeling, Pallet Labeling, Shrink Wrapping, and White Boxing.
Vertical Carousel	W	See Carousel.
Vertical Clearance	W	The distance between the top of a stack and the bottom of obstacles on the veiling of a facility, such as beams, trusses, and sprinklers. (KA)
Very Narrow Aisle	W	Very Narrow Aisle. A warehouse aisle that is purposed designed to be narrow. Normally this type of aisle has a wire imbedded in the center of floor, so that a forklift can sense the wire and stay precisely centered on the wire as it moves down the aisle.

Suggest New Words http://www.IDII.com/addword.htm

© 2004 Industrial Data & Information Inc.

Glossary of Supply Chain Terminology

Word	Type	Definition of Word
VNA	W	Very Narrow Aisle.
Wall Bumpers	W	Concrete-filled pipes 12 to 18 inches tall located to the side of the dock opening to protect adjacent walls from the impact of a misaligned truck trailer. (KA)
Wall-To-Wall Inventory	W	A full physical inventory count, which includes everything in the warehouse. (KA)
Warehouse	W	Place for receiving/storing goods and merchandise for hire. Warehouseman is bound to use ordinary diligence in preserving goods. (Cnf)
Warehouse Activity Report	W	A report that details all activities occurring within the warehouse facility, including merchandise arrivals, loading and unloading times and movements. AKA Activity Report. (KA)
Warehouse Management System	W	A management information system that controls warehouse activity, furnishing instructions to warehouse resources to manage operations. These systems typically interface with the Host, and RF devices that collect and disseminate information. Related software that may be included or tightly integrated with a WMS includes a Yard Management System, Parcel Manifesting System AKA Shipping Manifesting System, Slotting Optimization, Load Building AKA Load Optimization, and Transportation Management System.
Warehouse Receipt	W	A form that contains information describing the merchandise received into a warehouse. (KA)

Suggest New Words http://www.IDII.com/addword.htm

© 2004 Industrial Data & Information Inc.

Glossary of Supply Chain Terminology

Word	Type	Definition of Word
Warehouse Receipt	W	A receipt, usually non-negotiable, given for goods placed in a warehouse for storage. The receipt is a legal acknowledgement of responsibility of the care of goods. (KA)
Warehouse Within A Warehouse	W	A concept of taking subsets of the warehouse for a dedicated purpose. Each subset would be a "warehouse within a warehouse". E.G., By dedicating a zone (WWAW) for fast moving A items and another zone (WWAW) for B items, this would improve picking productivity.
Wave	W	A wave is a group of outbound orders to be released into picking together.
Wave Planning	W	Planning a wave of outbound orders by selection criteria. A person that plans the wave is called a planner. A planned wave may be one time or put on a regular schedule.
WC	Mfg W	Work Center.
Weather Seal	W	A rubber or canvas covering that extends out from a dock face to seal the gap between the dock and the trailer's entrance. (KA)
Weigh-In-Motion	W	A specialized weight scale that is imbedded in the conveyor line and connected to shipping software.
WIM	W	See Weigh-In-Motion.
WM	W	Warehouse Management. Inventory management & inventory control of product within facilities. See WMS
WMS	W	See Warehouse Management Systems.
WWAW	W	See Warehouse Within A Warehouse.
Zero Defects	Mfg W	A long-range objective that strives for defect-free products. (KA)

Suggest New Words http://www.IDII.com/addword.htm

© 2004 Industrial Data & Information Inc.

Glossary of Supply Chain Terminology

Word	Type	Definition of Word
Zone	W	A section of the warehouse that has exact characteristics based on material handling types and/or inventory management requirements. The Zone is the highest order in the location definition of: Zone, Aisle, Bay, Level, and Position.
Zone Pick	W	Each worker is to pick in a specific area, or zone, of the warehouse. Many warehouse management systems can be configured to zone picking, where workers are limited to one or a small number of zones.
Zone Picking	W	See Zone Pick.
Z Picking	W	A picking methodology where 1 picker proceeds down an aisle and picks from both rows and continues though the aisle for additional picks in that aisle. By watching the picker, one would view a Z travel path through the aisle if there were 4 bins to pick from. For other methods of picking, see Serpentine Picking, Zone Picking, Pick-to-Light, Pick-to-Carton, Batch Pick, Wave Planning, and Pick-and-Pass.

Suggest New Words   http://www.IDII.com/addword.htm

© 2004 Industrial Data & Information Inc.

Main Glossary

Full List of Acronyms, Keywords, & Abbreviations in the Supply Chain

Suggest New Words http://www.IDII.com/addword.htm

© 2004 Industrial Data & Information Inc.

Glossary of Supply Chain Terminology

Word	Type	Definition of Word
2 of 5	Std	Barcode Symbology.
2D Barcode	Std	Two Dimensional Barcode. A barcode that encodes information in two directions (X and Y). See 2D barcode symbology such as PDF 417, Datamatix, QRcode, or Maxicode.
2PC	C	Two Phase Commit
3 of 9	Std	Barcode with a variable length for alphanumeric codes. Commonly utilized in industry. No check digits utilized in 3 of 9 barcodes. Also known as Code 39.
3PF	Org	Third Party Fulfillment provider.
3PL	Org	Third Party Logistics provider. A separate company that manages the inventory and logistics of shipments. 3PLs provide transportation management, freight forwarding, customs brokerage, warehousing, kitting, light manufacturing and distribution services. A 3PL can be either asset or non-asset based. Many operations are outsourced to 3PLs.
4PL	Org	Fourth Party Logistics provider. A master contractor who manages an entire outsourced logistics network for a company. 4PLs traits include multi-modal services, advanced technology, international reach, and able to manage complex requirements. AKA as a LLP.
16K	Std	Barcode Symbology.
104	EDI	Air Shipment Information
110	EDI	Air Freight Details and Invoice
120	EDI	Vehicle Shipping Order
121	EDI	Vehicle Service
125	EDI	Multilevel Railcar Load Details

Suggest New Words   http://www.IDII.com/addword.htm

© 2004 Industrial Data & Information Inc.

Glossary of Supply Chain Terminology

Word	Type	Definition of Word
126	EDI	Vehicle Application Advice
127	EDI	Vehicle Buying Order
128	EDI	Dealer Information
129	EDI	Vehicle Carrier Rate Update
160	EDI	Transportation Automatic Equipment Identification
161	EDI	Train Sheet
163	EDI	Appointment Schedule Information
204	EDI	Motor Carrier Shipment Information
210	EDI	Motor Carrier Freight Details and Invoice
213	EDI	Motor Carrier Shipment Status Inquiry
214	EDI	Transportation Carrier Shipment Status Message
217	EDI	Motor Carrier Loading and Route Guide
218	EDI	Motor Carrier Tariff Information
250	EDI	Purchase Order Shipment Management Document
300	EDI	Reservation (Booking Request) (Ocean)
301	EDI	Confirmation (Ocean)
303	EDI	Booking Cancellation (Ocean)
304	EDI	Shipping Instructions
309	EDI	U.S. Customs Manifest
310	EDI	Freight Receipt and Invoice (Ocean)
311	EDI	Canadian Customs Information
312	EDI	Arrival Notice (Ocean)
313	EDI	Shipment Status Inquiry (Ocean)
315	EDI	Status Details (Ocean)
317	EDI	Delivery/Pickup Order
319	EDI	Terminal Information
322	EDI	Terminal Operations and Intermodal Ramp Activity
323	EDI	Vessel Schedule and Itinerary (Ocean)
324	EDI	Vessel Stow Plan (Ocean)
325	EDI	Consolidation of Goods in Container
326	EDI	Consignment Summary List
350	EDI	U.S. Customs Release Information

Suggest New Words   http://www.IDII.com/addword.htm

© 2004 Industrial Data & Information Inc.

Glossary of Supply Chain Terminology

Word	Type	Definition of Word
352	EDI	U.S. Customs Carrier General Order Status
353	EDI	U.S. Customs Events Advisory Details
354	EDI	U.S. Customs Automated Manifest Archive Status
355	EDI	U.S. Customs Manifest Acceptance/Rejection
356	EDI	U.S. Customs Permit to Transfer Request
357	EDI	U.S. Customs In-Bond Information
358	EDI	U.S. Customs Consist Information
361	EDI	Carrier Interchange Agreement (Ocean)
3-Tier	C	See N-Tier.
400	C	Web browser error for "Bad File Request". Check the URL and correct it. Web page or file may not exist.
401	C	Web browser error for "Unauthorized". Failed security authentication. Server requires login and password.
403	C	Web browser error for "Forbidden/Access Denied". Password is required to access this web page, directory, or entire site. See also 401.
404	C	Web browser error for "File not found". Could not find web page. Check typing of the URL string and retry. Site could be down or web page no longer exists.
404	EDI	Rail Carrier Shipment Information
408	C	Web browser error for "Request Timeout". Request not filled due to time limit exceeded or user canceled the request.
410	EDI	Rail Carrier Freight Details And Invoice
412	EDI	A16Trailer/Container Repair Billing
414	EDI	Rail Car-hire Settlements
417	EDI	Rail Carrier Waybill Interchange
418	EDI	Rail Advance Interchange Consist

Suggest New Words   http://www.IDII.com/addword.htm

© 2004 Industrial Data & Information Inc.

Glossary of Supply Chain Terminology

Word	Type	Definition of Word
419	EDI	Advance Car Disposition
420	EDI	Car Handling Information
421	EDI	Estimated Time of Arrival and Car Scheduling
422	EDI	Shipper's Car Order
423	EDI	Rail Industrial Switch List
425	EDI	Rail Waybill Request
426	EDI	Rail Revenue Waybill
429	EDI	Railroad Retirement Activity
431	EDI	Railroad Station Master File
432	EDI	Rail Description
433	EDI	Railroad Reciprocal Switch File
435	EDI	Standard Transportation Commodity Code Master
436	EDI	Locomotive Information
440	EDI	Shipment Weights
451	EDI	Railroad Event Report
452	EDI	Railroad Problem Log Inquiry or Advice
453	EDI	Railroad Service Commitment Advice
455	EDI	Railroad Parameter Trace Registration
456	EDI	Railroad Equipment Inquiry or Advice
460	EDI	Price Distribution or Response Format
463	EDI	Rail Rate Reply
466	EDI	Rate Request
468	EDI	Rate Docket Journal Log
475	EDI	Rail Route File Maintenance
485	EDI	Rate making Action
486	EDI	Rate Docket Expiration
490	EDI	Rate Group Definition
492	EDI	Miscellaneous Rates
494	EDI	Scale Rate Table
500	C	Web browser error for "Internal Error". Server problem, contact webmaster.
501	C	Web browser error for "Not Implemented". Server does not support the feature requested.

Suggest New Words http://www.IDII.com/addword.htm

© 2004 Industrial Data & Information Inc.

Glossary of Supply Chain Terminology

Word	Type	Definition of Word
502	C	Web browser error for "Service Temporarily Overloaded". Server is currently experiencing too many requests. Try again and try again later.
503	C	Web browser error for "Service Unavailable". Server is not available due to site is down, moved, too busy, or other issue.
503	EDI	Pricing History
504	EDI	Clauses and Provisions
601	EDI	Shipper's Export Declaration
602	EDI	Transportation Services Tender
622	EDI	Intermodal Ramp Activity
715	EDI	Intermodal Group Loading Plan
802.11	C	Standard Wireless Networking protocol. IT Directors must decide between 802.11a, 802.11b, and 802.11g.
802.11a	C Std	Wireless network standard that operates at 5 GHz with a maximum throughput of 54 Mbps. Range is 80 feet and supplies 12 separate non-overlapping channels. Good for high performance in a small, limited area. Incompatible with existing 802.11b and 802.11g equipment.
802.11b	C Std	Wireless network standard that operates at the crowded 2.4 GHz with a maximum throughput of 11 Mbps. Range is 328 feet and supplies 3 separate non-overlapping channels. Equipment is low cost and available. Compatible with existing 802.11g equipment.
802.11g	C Std	Wireless network standard that operates at the crowded 2.4 GHz with a maximum throughput of 54 Mbps. Range is 328 feet and supplies 3 separate non-overlapping channels. Interoperates with existing 802.11b equipment. Good for larger campuses.

Suggest New Words http://www.IDII.com/addword.htm

© 2004 Industrial Data & Information Inc.

Glossary of Supply Chain Terminology

Word	Type	Definition of Word
810	EDI	Invoice.
816	EDI	Orgal Relationships
820	EDI	Payment Order / Remittance Advice.
830	EDI	Planning Schedule With Release Capability
832	EDI	Price/Sales Catalog.
836	EDI	Contract Award.
840	EDI	Request For Quotation.
841	EDI	Specifications / Technical Information.
843	EDI	Response to Request for Quotation.
845	EDI	Price Authorization Acknowledgment/Status
846	EDI	Inventory Inquiry/Advice
847	EDI	Material Claim
850	EDI	Purchase Order.
851	EDI	Asset Schedule
852	EDI	Product Activity Data
853	EDI	Routing and Carrier Instruction
854	EDI	Shipment Delivery Discrepancy Information
855	EDI	Purchase Order Acknowledgment.
856	EDI	Ship Notice/Manifest. Advanced Ship Notice that lists goods that were shipped (per carton, pallet, container) and carrier information.
857	EDI	Shipment and Billing Notice
858	EDI	Shipment Information
859	EDI	Freight Invoice
860	EDI	Purchase Order Change Request - Buyer Initiated
861	EDI	Receiving Advice/Acceptance Certificate
862	EDI	Shipping Schedule
864	EDI	Text Message.
865	EDI	Purchase Order Change Acknowledgment/Request - Seller Initiated
866	EDI	Production Sequence

Suggest New Words http://www.IDII.com/addword.htm

© 2004 Industrial Data & Information Inc.

Glossary of Supply Chain Terminology

Word	Type	Definition of Word
867	EDI	Product Transfer and Resale Report
869	EDI	Order Status Inquiry
870	EDI	Order Status Report
871	EDI	Component Parts Content
875	EDI	Grocery Products Purchase Order
876	EDI	Grocery Products Purchase Order Change
878	EDI	Product Authorization/De-Authorization
879	EDI	Price Change
888	EDI	Item Maintenance
889	EDI	Promotion Announcement
891	EDI	Deduction Research Report
893	EDI	Item Information Request
894	EDI	Delivery/Return Base Record
895	EDI	Delivery/Return Acknowledgment or Adjustment
896	EDI	Product Dimension Maintenance
920	EDI	Loss or Damage Claim - General Commodities
924	EDI	Loss or Damage Claim - Motor Vehicle
925	EDI	Claim Tracer
926	EDI	Claim Status Report and Tracer Reply
928	EDI	Automotive Inspection Detail
940	EDI	Warehouse Shipping Order
943	EDI	Warehouse Stock Transfer Shipment Advice
944	EDI	Warehouse Stock Transfer Receipt Advice
945	EDI	Warehouse Shipping Advice
947	EDI	Warehouse Inventory Adjustment Advice
980	EDI	Functional Group Totals
990	EDI	Response to a Load Tender
997	EDI	Functional Acknowledgment. Transaction acknowledgment that trading partner received transaction on this date & time.
998	EDI	Set Cancellation

Suggest New Words http://www.IDII.com/addword.htm

© 2004 Industrial Data & Information Inc.

Glossary of Supply Chain Terminology

Word	Type	Definition of Word
- A -		**- A -**
A2A	B C	Application to Application. Connecting software application to another software application. See also EAI, B2B, XML, and EDI.
AAIA	Org	Automotive Aftermarket Industry Association (AAIA). See http://www.aftermarket.org/
AAPA	Org	American Association of Port Authorities. See http://www.aapa-ports.org/
AAR	Org	American Association of Railroads. See http://www.aar.org/
Abandonment	Tr	1. Proceeding where a carrier seeks authorization to stop service over all or part of its route/line, or to give up ownership/control of cargo or vessel. Must be approved by the ICC in the case of motor or rail proceedings. 2. Shipper or consignee relinquishes damaged freight carrier or refuses to accept delivery. 3. The act or relinquishing title to damaged or lost property to claim a total loss. (Cnf)
ABC	B	Activity Based Costing.
ABC	W	ABC classification. A user defined strategy for grouping parts within the warehouse (or organization). The grouping strategy is usually by product velocity, where fast movers are A items.
ABI	Tr	Automated Broker Interface.
ABM	W	Activity Based Management

Suggest New Words http://www.IDII.com/addword.htm

© 2004 Industrial Data & Information Inc.

Glossary of Supply Chain Terminology

Word	Type	Definition of Word
Absolute Liability	Tr	Condition in which a carrier is responsible for all liability and is not protected by normal exemptions found in a bill of lading or common law liability. (Cnf)
Acceptance	Tr	1. Acknowledged receipt by consignee of a shipment, terminating the common carrier contract. 2. A promise to pay, usually evidenced by inscribing across the face of the bill "accepted," followed by the date, place payable, and acceptor's signature. (Cnf)
Accessorial Charges	Tr	Charges for supplementary services and privileges (Accessorial Services) provided in connection with line-haul transportation of goods. These charges are not included in the freight charge and usually take the form of a flat fee. Some examples are: pickup/delivery, in-transit privileges, demurrage, switching, loading/unloading. (Cnf)
Accessorial Services	T	A service rendered by a carrier in addition to regular transportation service. (Cnf)
Account	B	1) An individual, institution, or other organization that purchases a company's products. 2) The general category of service as listed on the company books. (KA)
Accumulating Conveyor	Conv	Any conveyor designed to allow collection (accumulation) of material. May be roller, live roller, belt, or gravity conveyors. (Fs)

Suggest New Words   http://www.IDII.com/addword.htm

© 2004 Industrial Data & Information Inc.

Glossary of Supply Chain Terminology

Word	Type	Definition of Word
Accuracy Level	Uom	The percent of items located during an audit or sampling program that match the book inventory. The opposite measurement is the *error rate*. The accuracy rate can also be viewed as 100% minus the error rate. A company with an accuracy level of 95% will have an error rate of 5%. (KA)
ACF	Uom	Attainable Cubic Feet
Act of God	B	An unavoidable occurrence or accident produced by physical cause such as floods, earthquakes, most fires, or other natural disasters. (KA)
Acute Effect	MSDS	Health effects that usually occur rapidly, as a result of short-term exposure. (HC)
Acute Toxicity	MSDS	**Acute effects resulting from a single dose of, or exposure to, a substance. (HC)**
ADA	B	Americans with Disabilities Act. A 1990 US federal law that requires public facilities to be accessible to persons with disabilities.
ADA	Gov Tr	Airline Deregulation Act of 1978. (WA)
ADC	C	Automated Data Collection.
ADDR	B	Address. Common abbreviation used in business and in computer programming.
Ad Hoc	B	Ad Hoc Queries – An "on the spot" database query that is created by a user.
ADJ	B	Adjustment. Common abbreviation used in business and in computer programming.
Ad Valorem	Tr	A Latin phrase meaning "according to value." Freight rates set at a certain fixed percentage of the value of articles are known as ad valorem rates. (Cnf)

Suggest New Words http://www.IDII.com/addword.htm

© 2004 Industrial Data & Information Inc.

Glossary of Supply Chain Terminology

Word	Type	Definition of Word
Advanced Charge	Tr	Freight or charge on a shipment that is advanced by one transportation company to another, or to the shipper, to be collected from the consignee. (Cnf)
Advance Shipping Notification	C	Abbreviated ASN. Details on products being shipped with carton id numbers. May be transmitted by EDI, XML, and/or data on a barcode/RFID tag. See 856 for EDI format.
Advice of Shipment	Tr	Notice to local or foreign buyer that shipment has occurred, with packing and routing details. A copy of invoice usually is enclosed, and sometimes a copy of the bill of lading. (Cnf)
AES	C Std	Advanced Encryption Standard. See also DES & WEP.
AES	Tr	Automated Export System.
AESTIR	Tr	Automated Export System Trade Interface Requirements
Affinity	W	Affinity code. A user defined code for products that look alike. E.G., 10" and 12" adjustable chrome wrenches.
AFIF	Org	Association of Floral Importers of Florida. See http://www.siteblazer.net/afif
AGTM	Tr	Automated Global Trade Management. A computer & transportation term for software for advanced regulatory content management and trade automation services.
AGV	W	Automated Guided Vehicle.
AG/VS	W	Automated Guided Vehicle System.
AHTD	Org	Association for High Technology Distribution. See http://www.ahtd.org/
AI	C	Artificial Intelligence. Specialized computer software to rationally make educated decisions.
AIAG	Org	Automotive Industry Action Group. See http://www.aiag.org

Suggest New Words   http://www.IDII.com/addword.htm

© 2004 Industrial Data & Information Inc.

Glossary of Supply Chain Terminology

Word	Type	Definition of Word
AIDC	C	Automatic Identification and Data Control. See AIM.
AIM	Org	Automatic Identification Manufacturers Association. Industry association of the automatic identification and data capture (AIDC) industry. Their mission is to educate end users on AIDC solutions using Bar Codes, RFID, RF Data Communications, and much more. See http://www.aimusa.org
Air Waybill	Tr	A non-negotiable bill of lading that covers both domestic and international flights transporting goods to a specified destination. (WA) Abbreviated AWB.
Aisle	W	A corridor or passageway within a warehouse, which allows access to, stored goods. (KA)
Aisle Space Percentage	W	Abbreviated ASP. The proportion of a warehouse devoted to aisles. (KA)
AKA	B	Also Known As.
Algorithm	B	A method or process, such as an equation or computation, for solving a problem. (KA)
Alligator Lacing	Conv	Lacing attached to the belt with a hammer. (Fs)
All-Commodity Rate	Tr	Usually a carload/truckload rate that applies to multiple shipments that move at one time in one vehicle from the consignor to one consignee. An all-commodity rate is established based on actual transportation cost rather than "value of service." (Cnf)
Allocated Stock	W	Inventory that has been reserved, but not yet picked from stock, and thus is not available for other purposes.

Suggest New Words http://www.IDII.com/addword.htm

© 2004 Industrial Data & Information Inc.

Glossary of Supply Chain Terminology

Word	Type	Definition of Word
Allocation	W	Allocating inventory to a specific order. Order may be an outbound order (sales order, transfer order, work order, etc.). See Hard Allocation and Soft Allocation.
Allowance	Tr	Deduction from the weight or value of goods. Allowed if a carrier fails to provide necessary equipment and that equipment is furnished by the shipper. (Cnf)
Alongside	Tr	Point of delivery beside a vessel; statement designating where the title to goods passes from one party to another. (Cnf)
Alpha Factor	Pu	In exponential smoothing forecasting, the smoothing constant applied to the most recent forecast error. (KA)
Alphanumeric	C	A character set that contains letters, numbers, and other groups of symbols. (KA)
Alternate Commodity	W	A synonym for a Commodity, used by someone other than the owner of the product (i.e. the vendor's commodity name)
Alternate Routing	Tr	Routing that is less desirable than the normal, but results in identical terms. (Cnf)
Alternative Terms	Tr	The terms that a rail carrier offers in place of its full liability terms. The alternative terms are cheaper and carry less than a full limitation of liability. (WA)
AMP	B	Advertising, Marketing and Promotion (EM)
AMR	Org	AMR Research. A research Org for business applications & technology research. Website for AMR Research is http://www.amrresearch.com.

Suggest New Words http://www.IDII.com/addword.htm

© 2004 Industrial Data & Information Inc.

Glossary of Supply Chain Terminology

Word	Type	Definition of Word
AMS	Tr	Automated Manifest System. US Customs Automated Manifest System (AMS). TMS and SMS solutions can become AMS certified. Being AMS certified enables users of that software solution to submit advance cargo manifest information via the Automated Manifest System (AMS). This ensures that shipments won't face penalties, unnecessary delays and disruption to their supply chain.
AMT	B	Amount. Common abbreviation used in business and in computer programming.
ANLA	Org	American Nursery & Landscape Association. See http://www.anla.org/
Anniversary Billing	W	A method of public warehouse billing for storage in which the customer is billed a one-month storage charge for all products as they are received. The same unit, if still in storage, is billed an additional monthly charge on each monthly anniversary date thereafter. This method does not involve any pro-rating of time in storage and so requires that anniversary dates for each item in storage be separately identified. See split-month billing. (KA)
Annual Inventory	W	See Physical Inventory (KA)
Annular Nail	Pa	Pallet nail with annular (circular ring) threads rolled onto the shank. (P1)
ANOVA	W	Analysis of Variance
ANSI	Std	American National Standards Institute. ANSI provides information and conformity assessments to the US standards development community and industry. See http://www.ansi.org
AOQ	W	Average Outgoing Quality
AOV	W	Average Order Value

Suggest New Words http://www.IDII.com/addword.htm

© 2004 Industrial Data & Information Inc.

Glossary of Supply Chain Terminology

Word	Type	Definition of Word
AP	B	Accounts Payable. (1) The departments of a company responsible for verifying & payment of services rendered and products bought from vendors. (2) A software module utilized for entry of payable bills, aging of accounts payable, and printing of checks.
AP	C	Access Point. A computer hardware device with an antenna to permit wireless users (E.G., RF user) to roam freely among access points. The access point serves as a communications point for wireless clients and provides access to computer databases, electronic messaging, and so forth. Each access point has a connection to a wired LAN.
APDA	Org	American Parts Distributors Association, Inc. See http://www.apda.com
APEC	Org	Asian-Pacific Economic Caucus. See http://www.apec.org/
API	C	Application Programming Interface. A series of programs used to bring data in and out of the database. These programs are developed and maintained by the software application vendor to support interfacing their application to external applications.
APICS	Org	American Production & Inventory Control Society. Educational, non-profit association focused on Manufacturing, inventory, distribution, purchasing, logistics, and supply chain. Has CPIM and CIRM certification programs based upon APIC's body of knowledge. See http://www.apics.org

Suggest New Words http://www.IDII.com/addword.htm

© 2004 Industrial Data & Information Inc.

Glossary of Supply Chain Terminology

Word	Type	Definition of Word
Appearance	MSDS	A description of a substance (including color, size, and consistency) at normal room temperature and normal atmospheric conditions. (HC)
Application Package	C	A computer program or set of programs designed for a specific purpose. These can be custom designed, such as for inventory control or forecasting, or "off the shelf" such as spreadsheets or word processors. (KA)
APRA	Org	Automotive Parts Rebuilders Association. See http://www.apra.org/
Apron	W	The area directly outside the dock door upon which delivery vehicles can be parked or positioned for loading and unloading. This area is of the correct depth to allow trailer floors to line up with the warehouse floor. (KA)
APS	C	Advanced Planning and Scheduling software module. A Supply Chain Planning (SCP) tool to model supplier capacity, profiles, and products.
AR	B	Accounts Receivable. (1) The department of a company responsible for money due for services performed or merchandise sold on credit. (2) A software module utilized for invoicing, cash application, and aging of accounts receivables.
Arbitrary	Tr	1) Charge in addition to regular freight charge to compensate for unusual local conditions. 2) Fixed amount accepted by a carrier when dividing joint rates. (Cnf)
ARC	Org	IT research group with focus on Manufacturing, warehousing, and future growth trends. See http://www.arcweb.com

Suggest New Words http://www.IDII.com/addword.htm

© 2004 Industrial Data & Information Inc.

Glossary of Supply Chain Terminology

Word	Type	Definition of Word
Arguendo	Tr	A latin phrase meaning "for the sake of argument". (WA)
ARPA	Org	Advanced Research Projects Agency. See http://www.darpa.mil/
Arrival Notice	Tr	A notice, furnished to consignee, of the arrival of freight. (Cnf)
ASA	Org	American Supply Association. See http://www.asa.net/
ASCD	Org	Association of Service & Computer Dealers International. See http://www.ascdi.com/
ASCII	Std	American Standards Code for Information Interchange.
ASME	Org	American Society of Mechanical Engineers. See http://www.asme.org/
ASN	C	Advance Shipping Notification. Details on products being shipped with carton id numbers. May be transmitted by EDI, XML, and/or data on a barcode/RFID tag. See 856 for EDI format.
ASP	C	Application Service Provider. Provides application software on their computer systems.
ASP	W	Aisle Space Percentage.
ASPE	Org	American Society of Professional Estimators. See http://www.aspenational.com/
Asphyxiant	MSDS	A gas or vapor which can take up space in the air and reduce the concentration of oxygen available for breathing. Examples include acetylene, methane, and carbon dioxide. (HC)
ASQ	Org	American Society for Quality. See http://www.asq.org/
ASRS	W	See Automatic Storage & Retrieval System.
ASRS/AGVS	Org	ASRS/AGVS user's association. See http://www.asrs.org/

Suggest New Words http://www.IDII.com/addword.htm

© 2004 Industrial Data & Information Inc.

Glossary of Supply Chain Terminology

Word	Type	Definition of Word
Assembly Area	W	A warehouse location where materials, components, or finished products are collected and combined. (KA)
ASTL	Org	American Society of Transportation & Logistics. Promotes professionalism and continuing education in the field of transportation and logistics. See http://www.astl.org.
ASTM	Org	American Society for Testing Materials. Develops standards such as DSRC (dedicated short-range communications). See also ITS. See http://www.astm.org/
Astray Freight	Tr	Freight that has been separated from its freight bill. (Cnf)
Async	C	Abbreviation for Asynchronous.
ATA	Org	Air Transportation Association of America. See http://www.air-transport.org/
ATA	Org	American Trucking Association. See http://www.trucking.org/
ATA	Tr	Actual Time of Arrival. When the carrier's vehicle arrives at the agreed upon physical location.
ATD	Tr	Actual Time of Departure. When the carrier's vehicle departs from the agreed upon physical location.
ATMI	Org	American Textile Manufacturers Institute. See http://www.atmi.org/
ATO	W	Assemble To Order. A technique to postpone inventory that can be utilized as components and/or finished goods. Purpose is to reduce the inventory required on-hand but still keep service rate high, by assembling a finished goods product at the last moment.
ATP	B	See Available to Promise.

Suggest New Words http://www.IDII.com/addword.htm

© 2004 Industrial Data & Information Inc.

Glossary of Supply Chain Terminology

Word	Type	Definition of Word
ATRI	Org Tr	American Transportation Research Institute, formerly the ATA Foundation. See http://www.atri-online.org/
Auditing	Tr	Determining the correct transportation charges due the carrier, either before or after payment. Auditing involves checking the freight bill for extension errors, correct rate, commodity classification and description, weight, density, etc. Also called Freight Audit. (WA)
Audit Trail	B	The records and management controls that document business activities. Receipt, handling, and movement of materials throughout a warehouse are part of an audit trail. (KA)
Authority	Tr	Operating rights granted a motor carrier by the ICC. (Cnf)
Authorized Carrier	Tr	A person or company authorized by the ICC to transport goods as a common or contract carrier. (Cnf)
Auto-Ignition Temperature	MSDS	The temperature at which a material will ignite spontaneously or burn. (HC)
Automatic Storage & Retrieval System	W	Material Handling Equipment - specifically an automated, mechanized system for moving merchandise into storage locations and retrieving it from storage locations.
Available Stock	W	Inventory that is available to service immediate demand.
Available to Promise	B	Current inventory that is available & free from any commitments. This is the current on-hand inventory minus committed inventory for sales orders, transfer orders, Manufacturing orders (when product can also be a component), and vendor returns.

Suggest New Words http://www.IDII.com/addword.htm

© 2004 Industrial Data & Information Inc.

Glossary of Supply Chain Terminology

Word	Type	Definition of Word
AVDA	Org	American Veterinarian Distributors Association. See http://www.avda.net/
AVG	B	Average. Common abbreviation used in business and in computer programming.
AWA	Org	American Warehousemen's Association, which has been renamed to the IWLA.
AWB	Tr	See Air Waybill.
AWFS	Org	Association of Woodworking and Furnishings Suppliers. See http://www.awfs.org/
AWMA	Org	American Wholesale Marketers Association. See http://www.awmanet.org/
AYPI	B	And Your Point Is? AYPI is an acronym used in messaging other online users.
Axle	Conv	A non-rotating shaft on which wheels or rollers are mounted. (Fs)
- B -		**- B -**
B2B	B	Business-to-Business.
B2B2C	B	Business to Business to Consumer (EM)
B2C	B	Business-to-Consumer.
B2D	B	Business-to-Distributor.
B2E	B	Business-to-Employee.
B2G	B	Business-to-Government.
Back Haul	Tr	1. Return transportation movement, usually at less revenue than the original move. 2. Movement in the direction of lighter traffic flow when traffic generally is heavier in the opposite direction. 3. To move a shipment back over part of a route already traveled. (Cnf)

Suggest New Words http://www.IDII.com/addword.htm

© 2004 Industrial Data & Information Inc.

Glossary of Supply Chain Terminology

Word	Type	Definition of Word
Back Order	B	Items that have been ordered but cannot be shipped due to stockout. Merchandise on back order is scheduled for shipment when it becomes available. (KA)
Back Pressure	Conv	The amount of force applied to a package to stop the package or collection of packages. (Fs)
Bag Flattener	Conv	A mounting assembly used to hold one conveyor upside down over another conveyor in order to squeeze or flatten the product. (Fs)
BAL	B	Balance. Common abbreviation used in business and in computer programming.
Ball Table	Conv	A group of ball transfers over which flat surface objects may be moved in any direction. (Fs)
Ball Transfer	Conv	A device in which a larger ball is mounted and retained on a hemispherical face of small balls. (Fs)
Banding	W	Strapping to hold shipment together. Normally banding is plastic or metal strapping.
Banding Notch	Pa	See Strap Slot. (P1)
Bare Pulley	Conv	A pulley that does not have the surface of its face covered (or lagged). (Fs)
Barrell Truck, Barrell Wheeler	Tr	A dolly-like hand truck designed specifically to move drums or barrels. (Cnf)
Basing Point	Tr	Geographic point to which transportation rates are set so that rates to adjacent points can be constructed by adding to/deducting from the basing point rate. (Cnf)
Batch Pick, Batch Picking	W	A method of picking a single SKU to be distributed across multiple orders.

Suggest New Words   http://www.IDII.com/addword.htm

© 2004 Industrial Data & Information Inc.

Glossary of Supply Chain Terminology

Word	Type	Definition of Word
Baud	C	A measurement of the signaling speed of a data transmission device; equivalent to the maximum number of signaling elements, or symbols, per second that are generated; may be different from bit/second rat, however, especially at higher speeds, as several bits may be encoded per symbol, or baud with advance encoding techniques such as phase-shift keying.
Bay	W	A designated area within a section of a storage area outlined by markings on columns, posts, or floor. (KA)
Bay Storage	W	The use of a large designated area for storing merchandise. (KA)
BBS	W	Business Balanced Scorecard
BBSI	Org	Beauty & Barber Supply Institute. See http://www.bbsi.org/
Bearing	Conv	A machine part in or on which a shaft, axle, pin or other part rotates. (Fs)
Bed	Conv	That part of a conveyor upon which the load rests or slides while being conveyed. (Fs)
Bed Length	Conv	Length of bed sections only required to make up conveyor excluding pulleys, etc. that may be assembled at ends. (Fs)
Bed Width	Conv	Refers to the overall width of the bed section. (Fs)
Beginning Inventory	W	The inventory count at the beginning of the current period. See Ending Inventory and Physical Inventory. (KA)
Belt	Conv	A flexible band placed around two or more pulleys for the purpose of transmitting motion, power or materials from one point to another. (Fs)

Suggest New Words http://www.IDII.com/addword.htm

© 2004 Industrial Data & Information Inc.

Glossary of Supply Chain Terminology

Word	Type	Definition of Word
Belt Conveyor	Conv	A moving belt designed to carry merchandise. Belt conveyors are used to move materials between facilities, or between floors of a facility. (KA)
Belt Scraper	Conv	A blade or brush caused to bear against the moving conveyor belt for the purpose of removing material sticking to the conveyor belt. (Fs)
Belt Speed	Conv	The length of belt, which passes a fixed point within a given time. It is usually expressed in terms of 'feet per minute'. (Fs)
Between Rail Width	Conv	The distance between the conveyor frame rails on a roller bed, live roller or gravity type conveyor. Abbreviated BR. Also referred to as (BF) Between Frame. (Fs)
BF	Conv	See Between Frame or Between Rail Width.
BFN	B	Bye For Now. BFN is an acronym used in messaging other online users.
BI	B	Business Intelligence. From a computer savvy person, business intelligence is mining data from database(s). For a marketing & sales viewpoint, business intelligence may be both competitive information gathering as well as internal information gathering. Other terms similar to this are BPM, EPM, EIS, and OLAP.
Billed Weight	Tr	The weight shown on a freight bill. (Cnf)

Suggest New Words http://www.IDII.com/addword.htm

© 2004 Industrial Data & Information Inc.

Glossary of Supply Chain Terminology

Word	Type	Definition of Word
Bill of Lading	Tr	Bills of lading are contracts between the owner of the goods and the carrier. There are two types: A straight bill of lading in not negotiable. A negotiable or shipper's order bill of lading can be bought, sold or traded while goods are in transit and is used for letter of credits transactions. The customer usually needs the original or a copy as proof of ownership to take possession after payment for the goods. (WA) Abbreviated BOL, BL, B/L.
Bill of Materials	Mfg	A listing of components, parts, and other items needed to manufacture (or assemble) a product.
Bin	Pa W	Storage space for inventory at rest or being staged. A bin may be a physical bin or a virtual bin.
BIOS	C	Basic Input Output System
Birdyback	Tr	Moving highway freight by air. (Cnf)
Bisync	C	Abbreviation for Bisynchronous.
BizTalk	C	A XML standard for business commerce promoted & supported by Microsoft. Other XML standards include OAGIS by OAG and Rosetta Net.
BL, B/L	Tr	See Bill of Lading.
Blind Side	Tr	Right side of truck and trailer. (Cnf)
Block	Pa	Rectangular, square or cylindrical deck spacer, often identified by its location within the pallet as corner block, end block, edge block, inner block, center or middle block. (P1)
Blocking	Tr	Wood or metal supports to keep shipments in place in trailers. (Cnf)
Block Pallet	Pa	A type of pallet with blocks between the pallet decks or beneath the top deck. (P1)

Suggest New Words http://www.IDII.com/addword.htm

© 2004 Industrial Data & Information Inc.

Glossary of Supply Chain Terminology

Word	Type	Definition of Word
Block Pattern	W	A method of storing merchandise on a pallet in a pattern to allow a stable pallet load. (KA)
Bluetooth	C	A wireless protocol for transferring data & information between two devices.
BOA	Gov	Basic Order Agreement.
BOB	C	Best of Breed. Specialized software that has advanced functionality and utilizes "best practices".
Bob Tail	Tr	Tractor operating without a trailer. (Cnf)
Bogey	Tr	A two-axle assembly at the rear of some trailers or tractors. Also called a tandem axle. (Cnf)
Boilerplate	B	An agreement, purchase order, or other document that includes standard terms and procedures. (KA)
Boiling Point	MSDS	Temperature at which a liquid changes to a vapor state at a given pressure (usually sea level pressure = 760 mmHg). (HC)
BOL	Tr	See Bill of Lading.
BOM	Mfg	Bill of Material.
Bonded Warehouse	W	1. A warehouse approved by the Treasury Department and under bond/guarantee for observance of revenue laws. 2. Used for storing goods until duty is paid or goods are released in some other proper manner. (Cnf)
Bookland	Std	Barcode symbology used on books to encode ISBN on books.
Booster Conveyor	Conv	Any type of powered conveyor used to regain elevation lost in gravity roller or wheel conveyor lines. (Fs)
Bottleneck	B W	Congestion or significant slow down at an area due to an inefficient resource or process.

Suggest New Words http://www.IDII.com/addword.htm

© 2004 Industrial Data & Information Inc.

Glossary of Supply Chain Terminology

Word	Type	Definition of Word
Bottom Deck	Pa	Assembly of deckboards comprising the lower, load-bearing surface of the pallet. (P1)
Bottom Freight	Tr	Heavy freight that must be loaded on the trailer floor and not on top of other merchandise. (Cnf)
Box	Tr	Slang term for a trailer or container for ocean carriers. Slang term for a truck transmission. (Cnf)
BP	B	Best Practice (EM)
BP	B	Business Process.
BPI	B	Business Process Improvement.
BPM	B	Business Performance Management. Other terms similar to this are BI, EPM, EIS, and OLAP.
BPO	B	Business Process Optimization. AKA process re-engineering.
BPO	B	Business Process Outsourcing.
BPR	B	Business Process Reengineering.
BPS	C	Bits Per Second. A data transmission rate normally associated with modems and other handheld devices. See Baud Rate.
BR	Conv	See Between Rail Width.
Bracing	Tr	See Blocking.
Brake Motor	Conv	A device usually mounted on a motor shaft between motor and reducer with means to engage automatically when the electric current is cut off or fails. (Fs)
Brake Rollers	Conv	Air or mechanically operated brakes used underneath roller conveyor to slow down or stop packages being conveyed. (Fs)

Suggest New Words http://www.IDII.com/addword.htm

© 2004 Industrial Data & Information Inc.

Glossary of Supply Chain Terminology

Word	Type	Definition of Word
Break Bulk	Tr	The separation of a consolidated bulk load into smaller individual shipments for delivery to the ultimate consignee. The freight may be moved intact inside the trailer, or it may be interchanged and rehandled to connecting carriers. (WA)
Breakdown Time	Tr	A type of penalty pay, which is incurred when equipment breaks down. (Cnf)
Brick Pattern	W	A method of storing merchandise on a pallet in a pattern to accommodate items of unequal width or length. See Pinwheel Pattern. (KA)
Broker	Tr	1. An agent who arranges interstate movement of goods by other carriers. 2. An arranger of exempt loads for owner-operators and/or carriers. 3. One who arranges the buying/selling of goods for a commission. 4. A person who leases owned equipment to a carrier. (Cnf)
BTO	W	Built to Order
BTS	B	Balance To Ship. The remaining quantity of an order to be shipped.
Buffer Stock	W	Safety Stock. The stock held to protect against the differences between forecast and actual consumption, and between expected and actual delivery times of procurement orders, to protect against stock outs during the replenishment cycle. In calculating safety stock, account is taken of such factors as service level, expected fluctuations of demand and likely variations in lead time. See lead time inventories.

Suggest New Words   http://www.IDII.com/addword.htm

© 2004 Industrial Data & Information Inc.

Glossary of Supply Chain Terminology

Word	Type	Definition of Word
Bulk Carrier	Tr	Vessel that carries bulk commodities such as petroleum, grain, or ore, which are not packaged, bundled, bottled, or otherwise packed. (Cnf)
Bulk Freight	Tr	Freight not in packages or containers such as wheat, petroleum, etc. (Cnf)
Bulkhead	Tr	1) An upright wall in a trailer or railcar that separates and stabilizes a load. 2) A cargo restraining partition in a vehicle or vessel. (Cnf)
Bulk Storage	W	1) Storage of merchandise and materials in large quantities, usually in original or shipping containers. 2) Storage of unpackaged commodities. (KA)
Bumpers	W	Pieces of rubber located at the floor level of a dock opening to cushion the building from truck trailer impacts. (KA)
Burden Rate	Mfg	A standard cost added to every production hour to cover overhead expenses. (KA)
Butt Coupling	Conv	Angles or plates designed to join conveyor sections together. (Fs)
Butted Deckboard	Pa	An inner deckboard placed tightly against an adjacent lead deckboard during pallet assembly. (P1)
Buyer's Right to Route	Tr	When a seller does not pay freight charges, the purchaser has a right to designate the route for shipment. Seller is responsible for following the buyer's instructions. Complete routing is permitted for rail shipments, but only for the first carrier in motor shipments. (Cnf)

Suggest New Words http://www.IDII.com/addword.htm

© 2004 Industrial Data & Information Inc.

Glossary of Supply Chain Terminology

- C -		- C -
C or Ceiling	MSDS	The maximum allowable human exposure limit for an airborne substance, not to be exceeded even momentarily. Examples: hydrogen chloride, chlorine, nitrogen dioxide, and some isocyanates have ceiling standards.
Cab	Tr	Driver's compartment of a truck or tractor. (Cnf)
Cabatoge	Tr	A federal law that requires coastal and intercoastal traffic to be carrier in U.S. built and registered ships. (WA)
Cable Seal	Tr	A heavy steel cable used to secure closed trailer doors. It can only be removed with heavy duty cable cutters. (Cnf)
Cab-Over	Tr	Truck or tractor with a substantial part of the driving cab located over the engine. (Cnf)
CAD	C	Computer Aided Design. Computer Aided Drafting.
CAD	Pa	Computer-aided-design software that allows the design of the "right" pallet at the best value. See Pallet Design System (PDS). (P1)
CAE	C	Computer Aided Engineering (EM)
CALC	B	Calculation. Common abbreviation used in business and in computer programming.
CALM	Org	Canadian Association of Logistics Management. Non-profit association of business professionals interested in improving their logistics management skills. CALM has annual conferences, education, training, and periodical. See http://www.calm.org
CAM	Mfg	Computer Aided Manufacturing.

Suggest New Words http://www.IDII.com/addword.htm

© 2004 Industrial Data & Information Inc.

Glossary of Supply Chain Terminology

Canopy	W	A covering over the area outside a dock door, used to prevent rain, ice, or snow from interfering with truck loading. (KA)
Cantilever Rack	W	A storage rack in which the shelves rest on arms that extend from center posts. (KA)
CAPA	B	Corrective And Preventive Action. A plan of actions to correct & resolve a problem.
Capacity	Conv	The number of pieces, volume, or weight of material that can be handled by a conveyor in a unit of time when operating at a given speed. (Fs)
Capacity Load	Tr	1) A trailer loaded to the legal weight limit. 2) A trailer loaded so that no additional piece of freight, equal to the size of the largest piece tendered, will fit into the trailer. (Cnf)
CAPSTAN	Tr	Computer-Aided Planned Stowage and Networking system.
Captive Pallet	Pa	A pallet intended for use within the confines of a single facility, system or ownership; not intended to be exchanged. (P1)
CAR	B	Capital Appropriation Request.
Carbotage	Tr	A federal law that requires coastal and intercoastal traffic to be carrier in U.S. built and registered ships. (WA)

Suggest New Words http://www.IDII.com/addword.htm

© 2004 Industrial Data & Information Inc.

Glossary of Supply Chain Terminology 268

Carcinogen	MSDS	A material that causes cancer. A chemical is considered to be a carcinogen, by OSHA regulation, if: It has been evaluated by the International Agency for Research on Cancer (IARC), and found to be a carcinogen or potential carcinogen; or It is listed as a carcinogen or potential carcinogen in the Annual Report on Carcinogens published by the National Toxicology Program (NTP); or It is regulated by OSHA as a carcinogen; or There is valid scientific evidence in man or animals demonstrating a cancer-causing potential. (HC)
Carload	Tr	1) Quantity of freight required to fill a railcar. 2) Specified quantity necessary to qualify a shipment for a carload rate. Abbreviated as C/L and CL. (Cnf)
Carmack Amendment	Gov Tr	A federal statute that codifies the common law principle that a carrier is liable for the full value of goods lost, damaged or delayed while in its possession unless the shipper agrees in writing to a lower rate in return for a lower limitation of liability offered by the carrier. (WA)

Suggest New Words http://www.IDII.com/addword.htm

© 2004 Industrial Data & Information Inc.

Glossary of Supply Chain Terminology

Carousel	W	Material Handling Equipment - specifically an automated delivery system to move product from storage to picker, without any travel on picker. Two types of carousels are on the market, horizontal carousels and vertical carousels. Carousels are most productive when pick lists are downloaded from the WMS into the carousel system and optimized for multiple orders being picked at one time.
Carrier	Tr	The commercial entity responsible for deliveries of shipment to the customer. Carriers may specialize in small packages, air cargo, ocean cargo, rail cargo, LTL, or full truck load (TL). See also common carrier, contract carrier, authorized carrier.
Carrier Liability	Tr	A common carrier is liable for all shipment loss, damage, or delay with the exception of that caused by an act of God, act of a public enemy, act of a public authority, act or omission of the shipper, or the inherent vice or nature of the goods. (WA)
Carrying Costs	B	The cost of holding inventory, including taxes, depreciation, cost of invested capital, and insurance. Expressed as a percentage of total inventory, carrying cost is used in calculating economic order quantities. (KA)
Cart	Tr	A four wheeled platform used to move several pieces of freight across the dock at one time. (Cnf)
Cartage	Tr	1) The charge for pickup/delivery of goods. 2) The act of moving goods (usually short distances). (Cnf)
Carte Blanche	Tr	Full discretionary power, unlimited authority. (WA)

Suggest New Words   http://www.IDII.com/addword.htm

© 2004 Industrial Data & Information Inc.

Glossary of Supply Chain Terminology

Carton	W	A protective outer case that contains products. E.G., carton, tote, pallet.
CASE	C	Computer Aided Software Engineering.
Case Mark	Tr	Information shown on the outside of a shipping carton, including destination and contents. (Cnf)
Cash Before Delivery	Tr	Seller assumes no risk and extends no credit because payment is received before shipment. Abbreviated CBD. (Cnf)
Cash On Delivery	Tr	Buyer pays carrier the price of goods when they are delivered; seller assumes risk of purchaser refusing to accept goods. Abbreviated COD. (Cnf)
CASM	Tr	Carrier Assessment Selection and Management (EM)
Casters	Conv	Wheels mounted in a fork (either rigid or swivel) used to support and make conveyors portable. (Fs)
CBD	Tr	See Cash Before Delivery.
CBP	Org	U.S. Bureau of Customs and Border Protection. Cargo manifest rules require shippers to notify the CBP via electronic means before bringing cargo into or send cargo out of the U.S. CBP continues to implement Container Security Initiative (CSI) at major ports around the world. See http://www.cbp.gov
C-Commerce	B	Collaborative Commerce.
CCRA	Org Tr	Canada Customs and Revenue Agency. See http://www.ccra-adrc.gc.ca/
CDL	Tr	Commercial Driver's License. In the US, the FMCSA is monitoring security for those applying for a CDL to transport hazardous materials.

Glossary of Supply Chain Terminology

Term	Cat	Definition
Ceiling Hangers	Conv	Lengths of steel rod, attached to the ceiling, from which conveyors may be supported to provide maximum utilization of floor space or when required height exceeds floor support capability. (Fs)
CEMA	Org	Conveyor Equipment Manufacturers Association. See http://www.cemanet.org for information. CEMA has a number of educational & technical publications on conveyors, terminology, and standards. CEMA & ANSI produce Standard Publication 102 called "Conveyor Terms & Definitions".
Center Drive	Conv	A drive assembly mounted underneath normally near the center of the conveyor, but may be placed anywhere in the conveyor length. Normally used in reversing or incline application. (Fs)
CEO	B	Chief Executive Officer. Principal individual responsible for the activities of the company.
CERT	Mfg	Certificate of Analysis produced by a laboratory that is testing the product to certify it is within specification.
Certificate of Compliance	Mfg	A certification that suppliers or services meet specified requirements. (KA)
C&F	Tr	Cost and Freight. Quoted price includes the cost of transportation only. Buyer is responsible for loss & damage of shipment and may insure it at their expense. See also CIF
CAF	Tr	Cost and Freight. Quoted price includes the cost of transportation only. Buyer is responsible for loss & damage of shipment and may insure it at their expense. See also CIF
C Face Drive	Conv	A motor and reducer combination where the two units are flanged and are coupled for connection to each other and have one out-put shaft. (Fs)

Suggest New Words http://www.IDII.com/addword.htm

© 2004 Industrial Data & Information Inc.

Glossary of Supply Chain Terminology

CFAR	B	Collaboration Forecasting and Replenishment. See also CPFR.
CFO	B	Chief Financial Officer. Principal individual responsible for handling funds, financial planning of the company, and financial records.
CFPIM	B Mfg	Certified Fellow in Production and Inventory Management. Advanced certification program for Manufacturing and inventory by APICS.
CFR	Gov	Code of Federal Regulations
CFR	W	Case Flow Rack.
CFS	Tr	See Container Freight Station.
CGI	C	Common Gateway Interface. A program to take information provided from a web page and process it. A programmer can write a CGI application in a number of different languages such as C, C++, Java, and Perl. Alternatives to CGI are Microsoft's ASP and Macromedia's Cold Fusion (CFM).
CGMP	Mfg	Current Good Manufacturing Practice. Government regulations. Good Manufacturing Practice regulations (GMPs) are used by pharmaceutical, medical device, and food manufacturers as they produce and test products that people use. See GMP. See http://www.cgmp.com
CGMPs	Mfg	See CGMP.
CGP	B	Cost of Goods Purchased (EM)
Chain	Conv	A series of links pivotally joined together to form a medium for conveying or transmitting motion or power. (Fs)
Chain Conveyor	Conv	Any type of conveyor in which one or more chains act as the conveying element. (Fs)
Chain Drive	Conv	A power transmission device employing a drive chain and sprockets. (Fs)

Glossary of Supply Chain Terminology

Chain Guard	Conv	A covering or protection for drive or conveyor chains for safety purposes. (Fs)
Chain Roller Conveyor	Conv	A conveyor in which the tread rollers have attached sprockets which are driven by chain. (Fs)
Chamfered Deckboard	Pa	Deckboards with edges of one or two faces beveled, either along the full or specified length of board or between the stringers or blocks, allowing easier entry of pallet jack wheels. (P1)
Chargeback	B	Form used for recording transactions involving vendor returns. (KA)
Check Digit	C	In bar codes or data processing, a character added to ensure accuracy. KA)
CHEP	Org Pa	CHEP founded in 1946 and is a multi-national company known primary for CHEP pallets that it rents. The abbreviation CHEP stands for "Commonwealth Handling Equipment Pool". CHEP is equipment pooling system provider. See http://www.chep.com.
Chock	Tr	A wooden, metal, or rubber wedge used to block the wheels of a trailer at the dock. Also used in trailers to keep floor freight from shifting. (Cnf)
Chronic Health Effects	MSDS	Either adverse health effects resulting from long-term exposure or persistent adverse health effects resulting from short-term exposure. (HC)
Chronic Toxicity	MSDS	Adverse (chronic) effects resulting from repeated doses of or exposures to a substance over a relatively prolonged period of time. Ordinarily used to denote effects in experimental animals. (HC)

Suggest New Words http://www.IDII.com/addword.htm

© 2004 Industrial Data & Information Inc.

Glossary of Supply Chain Terminology

CHSCN	Org	Canadian Healthcare Supply Chain Network. "Promoting safe and quality healthcare through the implementation of optimal supply chain management practices and systems" is the vision of the CHSCN. See also EHCR and ECR. See http://www.loginstitute.ca/health_1.html
Chute	Conv	A trough through which objects are lowered by gravity. Can either be a slider bed or roller/wheel bed. (Fs)
CI	B	Continuous Improvement.
CIA	Org	Central Intelligence Agency. See http://www.cia.gov
CICS	C	Customer Information Control System.
CIDX	C	Chemical Industry Data Exchange is a data exchange standard based utilizing XML. Used for buying, selling and delivery of chemicals, also known as "Chem eStandards".
CIF	Tr	Cost, Insurance, and Freight. Quoted price includes cost of the goods, insurance of the goods, and freight charges. See also C&F.
CIFFA	Org	Canadian International Freight Forwarders Association. See http://www.ciffa.com
CIM	Mfg	Computer Integrated Manufacturing.
C-Inventory	B	Collaborative Inventory Management.
CIO	B	Chief Information Officer. Principal individual responsible for computer activities of the company.
CIRM	B	Certified in Integrated Resource Management. Certification program by APICS.
CIS	B	Customer Information Systems. See CRM

Suggest New Words http://www.IDII.com/addword.htm

© 2004 Industrial Data & Information Inc.

Glossary of Supply Chain Terminology

Term	Type	Definition
CITT	Org	Canadian Institute of Traffic & Transportation. See http://www.citt.ca/
CKD	Tr	Complete Knock Down logistics. I.E., Logistics provider would transport complete cars or motors broken down in parts to be assembled elsewhere.
C/L, CL	Tr	See Carload
Claim	Tr	1. Demand on transportation company for payment due to loss/damage of freight during transit. 2. Demand on transportation company for refund on overcharge. 3. Demand by an individual/company to recover for loss under insurance policy. (Cnf)
Claimant	Tr	The party filing a claim or suit. (WA)
Classifications	Tr	An alphabetical listing of commodities, assigning a class or rating into which the commodity is placed, with the truckload minimum weight necessary for application of a truckload rate ; used primarily to determine less-than-truckload rates based on class rates. (WA)
Class Rate	Tr	The rate charged for commodities grouped according to similar shipping characteristics. Class Rate applies to numbered/lettered groups/classes of articles contained in the territorial rating column in classification schedules. (Cnf)
Cleat	Conv	An attachment fastened to the conveying surface to act as a pusher, support check or trip, etc. to help propel material, parts or packages along the normal path of conveyor travel. (Fs)

Suggest New Words http://www.IDII.com/addword.htm

© 2004 Industrial Data & Information Inc.

Glossary of Supply Chain Terminology

Term	Type	Definition
Cleated Belt	Conv	A belt having raised sections spaced uniformly to stabilize flow of material on belts operating on inclines. Cleats may be a part of the belt or fastened on. (Fs)
Client	C	Any computer connected on a network that requests services from another computer on the network. Normally the client is a PC and is retrieving data from the "server" (another computer) that has the primary database(s) on it.
Clipper Lacing	Conv	Lacing attached to the belt with a clipper-lacing machine. (Fs)
CLM	Org	Council of Logistics Management. Educational, non-profit association focused on distribution, logistics, and supply chain. See http://www.clm1.org.
CLO	B	Chief Logistics Officer.
CLP	B	Country List Price
Clutchbrake Drive	Conv	Drive used to disengage motor from reducer and stop conveyor immediately without stopping the motor or cutting the power. (Fs)
Clutch Drive	Conv	Drive used to disengage motor from reducer without stopping the motor or cutting the power. (Fs)
CMs	B	Contract Manufacturers (EM)
CNT	B	Count. Common abbreviation used in business and in computer programming.
CO	B	Company. Common abbreviation used in business and in computer programming.
CO	Gov	Contracting Officer.
COBOL	C	Common Business Oriented Language
COD	Tr	See Cash On Delivery.
Code 11	Std	Barcode Symbology.
Code 16K	Std	Barcode Symbology.
Code 39	Std	Barcode Symbology.

Suggest New Words   http://www.IDII.com/addword.htm

© 2004 Industrial Data & Information Inc.

Glossary of Supply Chain Terminology

Term	Type	Definition
Code 128	Std	Barcode symbology with variable length for alphanumeric codes. More compact and higher validation than Code 39.
Codeabar	Std	Barcode symbology that has been used by Federal Express, libraries, and other niche markets.
Codification	Tr	The process of compiling, arranging, and systematizing the laws of a given jurisdiction into an ordered code. (WA)
COFC	Tr	See Container on a Flat Car. (KA)
COGS	B	Cost Of Goods Sold (EM)
Collar	Pa	Collapsible wooden container or bin, which transforms a pallet into a box. (P1)
Collectively Made Rates	Tr	Class rates established collectively by carrier members of regional rate bureaus and rate conferences under a grant of immunity from federal antitrust laws. (WA)
Collect Shipment	Tr	Shipment where collection of freight charges/advances is made by delivering carrier from the consignee/receiver. (Cnf)
Co-Load	Tr	Two shipments from different terminals combined to ship as one load. (Cnf)
Co-Managed Inventory	W	A support arrangement similar to Vendor Managed Inventory but where replacement orders for the vendor-owned stock are agreed by the user before delivery replenishment cycle. In calculating safety stock, account is taken of such factors as service level, expected fluctuations of demand and likely variations in lead time.
Combination	Tr	Truck or tractor coupled to one or more trailers (including semi-trailers). (Cnf)

Suggest New Words http://www.IDII.com/addword.htm

© 2004 Industrial Data & Information Inc.

Glossary of Supply Chain Terminology

Term	Code	Definition
Combustible Liquid	MSDS	Any liquid having a flash point at or above 100 °F (37.8 °C), but below 200 °F (93.3 °C), except any mixture having components with flash points of 200 °F (93.3 °C) or higher, the total volume of which make up 99 per cent or more of the total volume of the mixture. (HC)
Commercial Invoice	B Tr	Itemized list issued by seller/exporter in foreign trade showing quantity, quality, description of goods, price, terms of sale, marks/numbers, weight, full name/address of purchaser, and date. (Cnf)
Commodity	B	Any article of commerce. Goods, merchandise.
Commodity	W	One of the major outputs of the manufacturing effort, also known as product codes. Within a warehouse, an item uniquely identified to a particular customer.
Commodity, Exempt	Tr	One that may be transported in interstate commerce without operating authority or published rates. (Cnf)
Commodity Rate	Tr	1. A special (usually lower) rate for specific types of goods (usually exempt commodities). 2. A rate lower than class rates, established to cover the movement of a specific customer's freight or for a specific group of customers. (Cnf)
Common Carriage	Tr	Carriage performed by a common carrier that is not under contract with the shipper. (WA)
Common Carrier	Tr	A for-hire carrier that holds itself out to serve the general public. (WA)

Suggest New Words http://www.IDII.com/addword.htm

© 2004 Industrial Data & Information Inc.

Glossary of Supply Chain Terminology

Term	Code	Definition
Common Name	MSDS	Any designation or identification such as code name, code number, trade name, brand name, or generic name used to identify a chemical other than by its chemical name. (HC)
Concealed Damage	Tr W	When goods in an apparently undamaged container are damaged. (Cnf)
Concurrence	TR	Document signed by carrier and filed with the ICC. Verifies carrier participates in rates published in a tariff by a given agent. (Cnf)
Conditions to Avoid	MSDS	Conditions encountered during handling or storage that could cause a substance to become unstable. (HC)
Consign	Tr	Send goods to a purchaser or an agent to sell. (Cnf)
Consigned Stock	B	Finished goods inventories, in the hands of agents or dealers which are still the property of the supplier. (KA)
Consignee	Tr	The person who receives goods that were shipped.
Consignment	B	A transaction in which the title to goods remains with the shipper (the consignor) until the buyer (the consignee) sells the goods. (KA)
Consignor	Tr	The sender of a freight shipment. This is often termed the Seller or Vendor.
Consolidating	Tr	Combining small shipments to obtain reduced freight rated for higher volume. (KA)
Consolidation	Tr	Combining less-than-carload or less-than-truckload shipments to make carload/truckload movements. (Cnf)
Consolidation Point	Tr	The point at which small shipments are combined and loaded for reshipment. (KA)
Consolidator	Tr	A company that specializes in providing consolidation services to shippers. (KA)

Suggest New Words http://www.IDII.com/addword.htm

© 2004 Industrial Data & Information Inc.

Glossary of Supply Chain Terminology

Term	Tag	Definition
Constant Speed Drive	Conv	A drive with no provisions for variable speed or a drive with the characteristics necessary to maintain a constant speed. (Fs)
Constructive Notice	Tr	A rule of law binding the shipper to the terms and conditions contained in a law or a filed tariff. (WA)
Consumable Stock	W	A classification of stock used to describe items that are totally consumed in use - e.g. gift wrap, packing materials, etc.
Container	Tr W	Anything in which articles are packed. A standardized box used to transport merchandise particularly in international commerce. Marine containers are typically 8'X8' with length of 10', 30', or 40'. These containers may be transloaded from rail cars or ships onto a truck frame and delivered to their final destination. (KA)
Container Freight Station	Tr	A physical location by the carrier for cargo to be packed or unpacked by the carrier.
Containerization	Tr W	1. Using box-like device to store, protect, and handle a number of packages as a unit of transit. 2. Shipping system based on large cargo-carrying containers that can be interchanged between trucks, trains, and ships without re-handling contents. (Cnf)
Container on a Flat Car	Tr	Abbreviated COFC. A trailer without chassis or intermodal container shipped on a railroad flat car. (KA)
Contract	Tr	Legally defined as an offer, acceptance and bargained for consideration, a contract may be oral or in writing, if there is a valid offer, acceptance, and something of value exchanged between the two parties. (WA)

Suggest New Words http://www.IDII.com/addword.htm

© 2004 Industrial Data & Information Inc.

Glossary of Supply Chain Terminology

Contract Carriage	Tr	Transportation under a bilateral contract between a shipper and a carrier for a continuing period of time. (WA)
Contract Carrier	Tr	Any carrier engaged in interstate transportation of persons/property by motor vehicle on a for-hire basis, but under continuing contract with one or a limited number of customers to meet specific needs of each customer. Contract Carriers must receive an authorization permit from the ICC. (Cnf) See Carrier.
Contract of Adhesion	Tr	A standard-form contract prepared by one party, to be signed by the party in a weaker position. (WA)
Contract Rates	Tr	Rates that are part of a total contract negotiated between shipper and a carrier. (Cnf)
Contract Warehouse	W	See also Third Party Warehouse, Public Warehouse, and Warehouse.
Conventional	Tr	Tractor with the engine in front of the cab. (Cnf)
Converging	Conv	A section of roller or wheel conveyor where two conveyors meet and merge into one conveyor. (Fs)
Conveying Surface	Conv	Normal working surface of the conveyor. (Fs)
COO	B	Chief Operating Officer.
Cookie	C	A small data file created by a Web site while one is browsing on the Internet and stored on your local PC. The browser stores the data in a text file (a human readable file). This data file is then sent back to the web site's server every time the browser requests a page from the web site. The main design of cookies is to identify users and customized web pages for these users.

Suggest New Words   http://www.IDII.com/addword.htm

© 2004 Industrial Data & Information Inc.

Glossary of Supply Chain Terminology

Term	Cat	Definition
Cooler	W	A refrigerated space that holds material above freezing but usually below 50°. (KA)
CORBA	Std	Common Object Request Broker Architecture.
Corrosive Material	MSDS	A liquid or solid that causes visible destruction or irreversible alteration in human skin tissue at the site of contact. (HC)
COS	B	Conditions Of Satisfaction. Specific criteria that measures the process or product, for final determination whether it was successful or not.
Cost, Insurance, and Freight	Tr	The basis for quotation by seller that indicates seller will pay insurance and freight charges to destination only. (Cnf)
Cost of Capital	B	The cost to invest or borrow capital, usually expressed in a yearly percentage that is based on such factors as the current interest rate, or against alternative investments such as government securities. (KA)
Cost of Goods Sold	B	Abbreviated COGS. An accounting calculation to determine the total cost of merchandise that has been sold during a specific period. It is calculated by adding the net cost of merchandise purchases to the beginning inventory and then subtracting the cost of ending inventory. (KA)
Cost-Pass Through	Pa	A cost-share system where the partial cost of a pallet is passed-through from the purchaser to the receiver of the pallet. (P1)

Suggest New Words http://www.IDII.com/addword.htm

© 2004 Industrial Data & Information Inc.

Glossary of Supply Chain Terminology

Term	Uom	Definition
Cost per square foot	Uom	Abbreviated CSF. A monetary cost per unit of area to measure the basic cost of operating a warehouse. It represents the costs of the physcial space of a facility and the activities that occur within. The cost components include all the costs of operating the warehouse. (KA)
Cost-Per-Trip	Pa	Average cost of pallet use for a single one-way trip. (P1)
COTR	Gov	Contracting Officer's Technical Representative.
COTS	B Gov	Commercial Off the Shelf.
Counseling	B	Providing feedback to employees on work-related issues such as technical advice, performance reviews, or even personal problems. (KA)
CPFR	Std	Collaborative Planning, Forecasting and Replenishment. VICS Group sponsored this standard to optimize a manufacturers production and delivery of their product. See http://www.cpfr.org
CPG	W	Consumer Product Goods or Consumer Packaged Goods.
CPI	B	Consumer Price Index.
CPI	C	Characters Per Inch. On a printer, how many characters can print per inch in a printed line. 10 CPI is normal.
CPIM	B Mfg	Certified in Production and Inventory Management. Certification program that is focused on inventory & Manufacturing by APICS.
CPIO	B	Chief Process Improvement Officer (EM)
CPO	B	Chief Privacy Officer.
CPS	B	Collaborative Planning Solutions. Planning software providing a supply chain plan for a week, month, or longer.
CPS	C	Characters Per Second

Suggest New Words   http://www.IDII.com/addword.htm

© 2004 Industrial Data & Information Inc.

Glossary of Supply Chain Terminology

CPU	C	Central Processing Unit. The CPU is the "brain" of the computer and does all of the processing & calculations. For PC's, Intel and AMD sell chips that are the CPU. The CPU is located on the main circuit board (the motherboard) in a PC.
Credit Memo	B	A form used for recording and processing transactions involving a credit to suppliers or buyers. (KA)
Credit Rules	Tr	Tariff rules whereby the carrier agrees to deliver goods without receiving payment prior to delivery. The rules stipulate when payment may be paid without incurring a penalty, and may impose a penalty for late payments if the rules are published in accordance with FMCSA regulations. (WA)
CRM	C	Customer Relationship Management software module. Manage customer information for personalized service. See also ERM, PRM, and SRM.
CRO	B	Chief Risk Officer. CROs are either on the board or an executive in the company to manage all risk within the company, whether it is operational, information technology, credit, market, or reputation risk. Some IT managers prefer to be reporting to the CRO rather than the CFO.
Cross Bracing	Conv	Rods and turnbuckles placed diagonally across roller bed or live roller type conveyors to aid in squaring frames, necessary for tracking purposes. (Fs)
Cross Dock	Tr W	Transfer of freight from one trailer to another at a terminal. (Cnf)
Cross-Docking	W	A method of fulfilling orders that the movement of goods directly from receiving dock to shipping dock, bypassing the putaway of inventory to stock.

Suggest New Words   http://www.IDII.com/addword.htm

© 2004 Industrial Data & Information Inc.

Glossary of Supply Chain Terminology

Term	Cat	Definition
Crossover	Conv	A short section of conveyor placed in a conveyor when drive is switched to opposite side of conveyor. (Fs)
Cross-training	W	Providing training or experience in several different warehousing tasks and functional specialties in order to provide backup workers. (KA)
Crowned Pulley	Conv	A pulley that tapers equally from both ends toward the center, the diameter being the greatest at the center. (Fs)
CRP	Mfg	Capacity Requirements Planning.
CRP	Mfg	Continuous Replenishment Program.
CRT	C	Cathode Ray Tube. A computer screen used for data entry. (KA)
CS	B	Customer Service.
CSF	Uom	Cost Per Square Foot.
CSI	Tr	U.S. Customs Container Security Initiative. Places US Customs inspectors overseas to identify high-risk shipments. See www.customs.ustreas.gov See also C-TPAT, FAST, and HSA.
CSO	B	Chief Security Officer.
CSR	B	Customer Service Representative. A person with heavy contact with the customer. May be dedicated to one or more customers and duties will vary dependent upon type of industry (3PL, Manufacturer, Distributor).
CSS	C	Cascading Style Sheets. Using cascading style sheets in web pages for a website, will save time & bring forth consistency for font, font size, font color, and more.
CSV	C	Comma Separated Values. A data file, where fields are separated ("delimited") by commas.
CTD	B	Capable To Deliver. Order commitment based upon available inventory quantities and pipeline (manufacturing or in-transit quantities).

Suggest New Words http://www.IDII.com/addword.htm

© 2004 Industrial Data & Information Inc.

Glossary of Supply Chain Terminology

Term	Cat	Definition
CTDA	Org	Ceramic Tile Distributors Association. See http://www.ctdahome.org/
CTO	B	Chief Technology Officer.
CTO	Mfg W	Configure To Order. See also ETO and MTS.
CTP	B	Capable to Promise.
CTP	W	Capable To Promise. See ATP / Available To Promise.
C-TPAT	Tr	Customs-Trade Partnership Against Terrorism. Requires importers, carriers and custom brokers to document security procedures. Accepted companies qualify for expedited processing and exemption from physical inspections. See www.customs.ustreas.gov. See also TSA, CSI, FAST, and HSA.
CTQ	B	Critical to Quality.
CTR	Mfg	Cycle Time Reduction
Cube Loading	Tr W	The process of loading merchandise onto pallets or other unit-loading techniques to allow several unit loads to be stacked for transportation. (KA)
Cube Rate	Tr	A rate based on trailer space instead of weight. Used for light, bulky loads. (Cnf)
Cube Utilization	Tr W	The percentage of space occupied compared to the space available. (KA)
Cubiscan	W	A device that cubes and weighs products that are placed unto it. Cubiscan is a registered Trademark ™ of Quantronix company. See http://www.cubiscan.com or website at http://www.quantronix.com.
CUL8R	B	See You Later. CUL8R is an acronym used in messaging other online users.
Cure Notice	Gov	Notice sent to a contractor that contract is in jeopardy and documentation is immediately requested that describes how the situation will be remedied.

Suggest New Words   http://www.IDII.com/addword.htm

© 2004 Industrial Data & Information Inc.

Glossary of Supply Chain Terminology

Term	Type	Definition
Curve Conveyor	Conv	Any skatewheel, roller or belt conveyor that is produced with a degree of bend so as to convey products away from the straight flow. (Fs)
CUSDEC	Tr	Customs Declaration Message. An EDI (UN/EDIFACT) transaction to send information from one party to another.
CUST	B	Customer. Common abbreviation used in business and in computer programming.
Customer Pickup	W	1) Merchandise picked up by the customer at the warehouse. 2) The act of picking up merchandise at the dock. AKA Will-Call or Walk-In. (KA)
Customer Service	B	The aspect of logistics which provides the highest level of service at the lowest cost. An all-encompassing term for non-price and non-physical product features of a product, usually consisting of service lead time, customer contact and problem resolution. (KA)
Customhouse Broker	Tr	A person or firm licensed to enter and clear goods through Customs. (WA)
Customs Bond	Tr	A contract between a principal, usually an importer, and a surety, which is obtained to insure performance of an obligation imposed by law or regulation. The bond covers potential loss of duties, taxes and penalties for specific types of transactions. (WA)
Customs Broker	Tr	A specialist in customs procedures that processes the entry and clearance of goods into the country for a fee.
Custom Software	C	Software specially developed for or by the user. (KA)
Customs Tariff	Tr	A schedule of charges assessed by the government on imports/exports. (Cnf)
CWA	Org	Central Wholesalers Association. See http://www.asaonline.org/
CWDA	Org	Canadian Wholesale Drug Association.

Suggest New Words http://www.IDII.com/addword.htm

© 2004 Industrial Data & Information Inc.

CWT	Uom	Hundredweight. (KA)
Cycle Counting	W	Cycle counting is the physical counting of stock on a perpetual basis, rather than counting stock periodically. A cycle is the time required to count all items in the inventory at least once. The frequency of cycle counting can be varied to focus management attention on the more valuable or important items or to match work processes. Some of the systems used are: ABC system with the highest count frequency for items with the highest annual usage value, Reorder system when stocks are counted at the time of order, Receiver system with counting when goods are received, Receiver system with counting when goods are received reached to confirm that no stock is held, Transaction system where stocks are counted after a specified number of transactions.

Glossary of Supply Chain Terminology

- D -		- D -
DAM	B	Digital Asset Management (EM)
Damage Claim	Tr	Request by a shipper or consignee for reimbursement from a carrier for damage to a shipment. (KA)
DARPA	Org	Defense Advance Research Project Administration. See http://www.darpa.mil/
Data Base	C	Abbreviated DB. Data stored in a form that allows for flexible sortation and report generation. Common data bases include customer lists, routes, carrier selection maps, rate files, and inventory files. (KA)
Datamatrix	Std	2D barcode symbology.
Date Code	Mfg W	A label showing the date of production. In the food industry, it becomes an integral part of the lot number. (KA)
DB	C	Data Base. E.G., Oracle, Sybase, Informix, SQL Server, Progress.
D&B	Org	Dun & Bradstreet. US based company providing financial statistics and credit ratings of businesses. See http://www.db.com.
DBA	C	Data Base Administrator. An IT person that is highly trained to keep the databases operational. Also responsible for backups and optimization of those databases.
DBMS	C	Data Base Management System. A computer application dedicated for the storage of data. The DBMS is organized into databases, tables, columns, rows, and indexes. In older terminology, columns were called fields and rows where called records. AKA DB.
DC	W	See Distribution Center.
DDE	C	Dynamic Data Exchange.

Suggest New Words http://www.IDII.com/addword.htm

© 2004 Industrial Data & Information Inc.

Glossary of Supply Chain Terminology

Term	Type	Definition
DE	B	Debt/Equity ratio (EM)
Dead Axle	Tr	Non-powered rear axle on tandem truck or tractor (also called "tag axle"). (Cnf)
Deadhead	Tr	1) A trailer moving empty. 2) A shipment moving without charges. 3) A ride-along driver. (Cnf)
Dead Stock	W	A product that does not move from its storage location for a long time. (KA)
Dead Weight Tonnage	Tr	Estimated number of tons of cargo a vessel can carry when loaded to maximum depth. (Cnf)
Deck Opening	Pa	The space between the deckboards of a pallet. (KA)
Declared Value	Tr	1) Assumed value of shipment unless shipper declares higher value. 2) Stating lower value on a shipment to get a lower rate. (Cnf)
Decline Conveyor	Conv	A conveyor transporting down a slope. (Fs)
Deck	Pa	One or more boards or panels comprising the top or bottom surface of the pallet. (P1)
Deckboard	Pa	Element or component of a pallet deck, oriented perpendicular to the stringer or stringerboard. (P1)
Deckboard Spacing	Pa	Distance between adjacent deckboards. (P1)
Deckboard Span	Pa	Distance between deckboard supports (stringers, stringerboards or blocks). (P1)
Deck Mat	Pa	Assembly of deckboards and stringerboards, forming the deck of a block pallet. (P1)
Decomposition	MSDS	Breakdown of a material or substance (by heat, chemical reaction, electrolysis, decay, or other processes) into simpler compounds. (HC)
Decomposition Products	MSDS	Describes hazardous materials produced during heated operations. (HC)

Suggest New Words   http://www.IDII.com/addword.htm

© 2004 Industrial Data & Information Inc.

Glossary of Supply Chain Terminology

Decree of Incline	Conv	Angle of slope (in degrees) that a conveyor is installed. (Fs)
Deductible	Tr	An amount of loss accepted by the insured as a condition for a reduction in the insurance premium or carrier's rate. (WA)
Deep-Lane Storage	W	Storage of merchandise greater than one unit deep on one or both sides of an aisle. See Double-Deep Lane Storage and Single Deep. (KA)
Deferred Rebate	Tr	Carrier returns a portion of freight charges to shipper. In exchange, shipper gives all/most shipments to carrier over specified period, usually six months. Rebate payment is deferred for similar period. (Cnf)
Deflection	Pa	The amount of deformation or bending in a pallet or pallet component under load. (P1)
Deflection	W	The sag, bend, or deformation of a rack beam platform or container side due to the weight of a load. (KA)
Dekitting, Dekit	Mfg W	When a work order has been canceled, the process to return the components back to inventory is called dekitting.
Delivered Price	B	A price for merchandise that includes transportation charges to a delivery point agreed upon by the seller and buyer. (KA)
Delivery Receipt	Tr	A carrier-prepared form that is signed by the consignee at the time of delivery. See Proof of Delivery (KA)
Delivery Window	Tr	A period during which a delivery (or deliveries) must be made. (KA)
Demand Forecasting	Pu	Determining predictions of a product's future usage using a judgmental approach, a relational approach, or a time series-approach. (KA)

Suggest New Words http://www.IDII.com/addword.htm

© 2004 Industrial Data & Information Inc.

Glossary of Supply Chain Terminology

Demurrage	Tr	1) Detention of a ship, freight car, or other cargo conveyance during loading or unloading beyond the scheduled time of departure. 2) Compensation paid for such detention.
Density	MSDS	The mass of a substance per unit volume. The density of a substance is usually compared to water, which has a density of 1. Substances which float on water have densities less than 1; substances which sink have densities greater than 1. (HC)
Density	Tr W	The weight of an article per cubic foot. (Cnf)
Department of Transportation	Org	The federal agency that regulates the highway transportation of freight, including commodities designated as hazardous material. See http://www.dot.gov/
Deregulation	B	The reduction of governmental control of business, especially to permit increased competition in a free market. (WA)
Dermal	MSDS	Used on or applied to the skin. (HC)
Dermal Toxicity	MSDS	Adverse effects resulting from skin exposure to a substance. (HC)
DES	Std	Data Encryption Standard. See also AES & WEP.
Detention	Tr	A charge made for a vehicle held by, or for, consignor or consignee for loading, unloading or for forwarding directions. (Cnf)
Dial-Up	C	A term used to indicate that computers are communicating though modems on standard telephone lines rather than via a dedicated communication link (such as a network or leased telephone lines). (KA)
Differential	Tr	Amount added to/deducted from base rate to make rate to/from some other point or via another route. (Cnf)

Suggest New Words http://www.IDII.com/addword.htm

© 2004 Industrial Data & Information Inc.

Glossary of Supply Chain Terminology

Differential Curve	Conv	A curved section of roller conveyor having a conveying surface of two or more concentric rows of rollers. Also referred to as a Split Roller design. (Fs)
Dimensions	Pa	See Pallet Dimensions. (P1)
DIR	W	Drive In Rack.
DISA	Org	Data Interchange Standards Association. See http://www.disa.org/
Discharge End	Conv	Location at which objects are removed form the conveyor. (Fs)
Discount	Tr	A reduction from the full amount of rate and charges offered by the carrier. (WA)
Discrimination	B	The effect of a statute or established practice that creates disadvantages for a certain group of persons or interests. (WA)
Discrimination	Tr	Difference in rates not justified by costs. (Cnf)
Dispatch	Tr	The process of combining a driver, a tractor, a trailer, and a load. (KA)
Distribution Center	W	A modern warehouse that processes inventory based upon the direction of the corporate systems. Abbreviated DC.
Distribution System	B	The system and processes of transporting goods within and among plants, warehouses, and other facilities. (KA)
Distributor	B	A business that is in the middle of a supply channel. Distributors buy and sell finished goods. They man alter, assemble, combine, or otherwise add value to the goods. (KA)
Diverging	Conv	A section of roller or wheel conveying which makes a connection for diverting articles from a main line to a branch. (Fs)

Suggest New Words http://www.IDII.com/addword.htm

© 2004 Industrial Data & Information Inc.

Glossary of Supply Chain Terminology

Diversification	B	Spreading a company's risk in case of a downturn in the demand for any given product or service. Diversification may involve expansion into different markets or product lines. (KA)
Diversion	Tr	A change made in consignee, destination, or shipment route while in transit. (Cnf)
DL	B	Direct Labor
DLL	C	Dynamic Link Library.
DMAIC	B	Define, Measure, Analyze, Improve, Control.
DOA	Tr	Dead On Arrival. Imperfect delivery.
DOC or DoC	Org	U.S. Department of Commerce. See http://www.doc.gov/
Dock	Tr W	The floor or platform where trucks load and unload. (Cnf)
Dockboard	W	A device for bridging the gap between the warehouse floor and a vehicle's load bed. (KA)
Dock Face	W	The outside wall of the dock door area. (KA)
Dock Fire Door	W	Generally used with enclosed apron areas, this is a safely feature to protect the interior dock area from fires that may occur on the apron or in the trailer itself. (KA)
Dock Leveler	W	A manually or hydraulically operated plate, located at the dock entrance, that can be raised and lowered approximately one foot to accommodate varying trailer floor heights. (KA)
Dock Light	W	A flood light positioned so it illuminates the interior of a trailer, which is not obstructing loading and unloading activities. (KA)
Dock Plate	W	A moveable metal ramp that allows access to a rail car or trailer. (KA)
Dock Receipt	W	A receipt issued for a shipment at a pier or dock. (KA)

Suggest New Words   http://www.IDII.com/addword.htm

© 2004 Industrial Data & Information Inc.

Glossary of Supply Chain Terminology

Term	Cat	Definition
DOD or DoD	Org	U.S. Department of Defense. See http://www.defenselink.mil/
DOE	Org	U.S. Department of Energy. See http://www.doe.gov and http://www.energy.gov.
DOL	Org	U.S. Department of Labor. See http://www.dol.gov.
Dolly	Tr W	A non-motorized, two-wheeled hand truck for moving freight around the dock. (Cnf)
DOS	C	Disk Operating System. One of the first operating systems for PC's was DOS by IBM & Microsoft. DOS is obsolete. Some computer geeks during those early PC years, called DOS another acronym, "Dumb Operating System". See Windows.
DOT	Tr	See Department of Transportation. See http://www.dot.gov/
Double-Deep Lane Storage	W	Rack storage of merchandise two loads deep on one or both sides of an aisle. See Deep-Lane Storage and Single Deep. (KA)
Double-Face Pallet	Pa	A pallet with top and bottom deckboards extending beyond the edges of the stringers or stringerboards. (P1)
Downtime	B	The time when equipment is scheduled for operational, but is idle for maintenance, repairs, or changeovers. (KA)
DPI	C	Dots Per Inch
DPM	B Mfg	Defects Per Million.
DPMO	Mfg	Defects per Million Opportunities
Draft Curtain	W	A fire curtain that hangs from the warehouse ceiling to reduce the spread of fire. (KA)
DRAM	C	Dynamic Random Access Memory
Drawback	B	Refund of customs duties paid on material imported and later exported. Also known as "Duty Drawback". (Cnf)

Suggest New Words http://www.IDII.com/addword.htm

© 2004 Industrial Data & Information Inc.

Glossary of Supply Chain Terminology

Term	Cat	Definition
Drayage	Tr	Transporting freight by truck, primarily in local cartage. (Cnf)
Drive	Conv	An assembly of the necessary structural, mechanical and electrical parts that provide the motive power for a conveyor, usually consisting of motor/reducer, chain, sprockets, guards, mounting base and hardware. (Fs)
Drive Axle	Tr	The axle(s) which are connected to the engine by a drive shaft and power the vehicle. Also called "power axle". (Cnf)
Drive-In Rack	W	Storage rack which provides side rails to allow high stacking in deep rows. Unlike drive-through rack, it provides access only from the aisle. (KA)
Drive Pulley	Conv	A pulley mounted on the drive shaft that transmits power to the belt with which it is in contact. Pulley is normally positive crowned and lagged. (Fs)
Drive Screw Nail	Pa	Helically (continuous spiral) threaded pallet nail. (P1)
Drive-Through Rack	W	Storage rack which provides side rails to allow high stacking of products in deep rows and access to the from either end of the row. (KA)
Driver Collect	Tr	Refers to a shipment for which the driver must collect freight charges at the time of delivery. (Cnf)
Driveway Installation	W	A ramp located on the outdoor apron of the dock, used to raise and lower a truck trailer, so that its floor becomes level with the dock floor. (KA)
DRP	Mfg	Distribution Requirements Planning or Distribution Resource Planning. The distributed version of MRP and MRP II.
Dry Chemical	MSDS	A powdered, fire-extinguishing agent usually composed of sodium bicarbonate, potassium bicarbonate, etc. (HC)
DSD	B W	Direct Store Delivery. See also FRM.

Suggest New Words http://www.IDII.com/addword.htm

© 2004 Industrial Data & Information Inc.

DSRC	Tr	Dedicated Short-Range Communications. See ASTM and ITS.
DSS	C	Decision Support System.
DT	B	Date. Common abbreviation used in business and in computer programming.
DTD	C	Document Type Definition. RosettaNet has developed an XML standard, using PIPs & DTDs.
DTS	W	Dock To Stock.
Dual Rate System	Tr	An international water carrier pricing system in which a shipper signing an exclusive use agreement with the conference pays a rate 10 to 15 percent lower than non-signing shippers pay for an identical shipment. (WA)
DUB	Tr	A 28-foot trailer designed to be pulled two or three at a time by one tractor. Also known as pup or doubles. (Cnf)
Dumb Terminal	C	See Green Screen.
Dumbwaiter	W	A miniature freight elevator with a car moved by a hand-operated pulley. It is used for carrying small parcels. (KA)
Dunnage	Tr	Term used for cardboard, empty pallets, plywood, foam rubber, air bags, or other items used to cushion or protect freight while in transit. (Cnf)
DUNS	B	Data Universal Numbering System. The Dun & Bradstreet code (DUNS Number) assigned to uniquely identify a company.
DuPont Model	B	A display of the factors of return on investment which pinpoints causes for increase or decrease in return. The model divides a business into three areas: financial management, asset management, and operations. Pretax return on investment (PROI) equals financial leverage times asset turnover times net profit margin. Originally developed by E.I. du Pont de Nemours. (KA)

Suggest New Words   http://www.IDII.com/addword.htm

© 2004 Industrial Data & Information Inc.

Glossary of Supply Chain Terminology

Dutchman	Conv	A short section of belt provided with lacing, in a conveyor belt that can be removed when take-up provision has been exceeded. (Fs)
Duty	B	A taxed assessed by the government for importing and exporting goods. (KA)
Duty Drawback	B	Refund of customs duties paid on material imported and later exported. Also known as Drawback. (Cnf)

Suggest New Words http://www.IDII.com/addword.htm

© 2004 Industrial Data & Information Inc.

Glossary of Supply Chain Terminology

- E -		- E -
EAI	C	Enterprise Application Integration. E.G., Interfacing data between ERP and WMS. Also, mapping between a WMS and TMS.
EAN	Std	European Article Numbering.
EAN	Std	International Article Numbering Association.
Earmarked Material	W	On-hand material that is reserved an physically identified rather than simply allocated in inventory controls. (KA)
EAS	B	Electronic Article Surveillance.
E-business, EB	B	Electronic Business.
ebXML	C	Electronic business eXtensible Mark-up Language. ebXML is an evolutionary framework to accommodate various types of e-transaction technologies ranging from EDI to XML. It is a joint initiative of the UN and OASIS, to advance electronic business by promoting open, collaborative development of interoperability specifications.
EC	Org	European Commission or European Community. See http://www.europa.eu.int/
EC, E-commerce	B	Electronic Commerce.
ECA	Org	Express Carriers Association. See http://www.expresscarriers.com/
ECCC	Org	Electronic Commerce Council of Canada. Associated with the UCC Council on promoting standards for commerce. See http://www.eccc.org
ECCN	Tr	Export Control Classification Number

Suggest New Words http://www.IDII.com/addword.htm

© 2004 Industrial Data & Information Inc.

Glossary of Supply Chain Terminology

Economic Life	Pa	Output from program that identifies the number of trips the pallet will make, provided it is properly repaired, which maximizes a return on investment. (P1)
ECR	B W	Efficient Customer Response.
EDI	C EDI	Data format specification. See the Electronic Data Interchange section in this glossary.
EDIFACT	Std EDI	EDI for Administration, Commerce and Transport. EDIFACT is n EDI standard for international usage and was developed by the United Nations.
EFF	B	Effective. Common abbreviation used in business and in computer programming.
EFQM	B	European Foundation for Quality Management (EFQM) Excellence Model.
EFT	B	Electronic Funds Transfer.
EFTA	Org	European Free Trade Association. See http://www.efta.int/
EHCR	B W	Efficient Healthcare Consumer Response. To realize supply chain cost savings through the adoption of EDI, bar-coding, and other strategies such as Activity Based Costing. See also ECR.
EHS	MSDS	Environmental Health and Safety Office.
EI	W	See Ending Inventory
EIC	Tr	Export Information Code
EIN	B	Employer Identification Number

Suggest New Words http://www.IDII.com/addword.htm

© 2004 Industrial Data & Information Inc.

Glossary of Supply Chain Terminology

EIP	C	Enterprise Information Portal. Software for an enterprise website(s) offering a user-friendly interface to gain access to enterprise data. EIPs combine search, content management, and DB technologies to deliver information tailored for each user. EIP solutions include BroadVision, Northern Light Technology, and others. EIP can draw from the ERP database, but will also draw information from it's own content management database. Therefore EIP is –not- an ERP system, but a content delivery system to automate common user information requests.
EIS	C	Executive Information System software module. Other terms similar to this are BI, BPM, EPM, and OLAP.
ELA	Org	European Logistics Association. A federation of 36 national organizations, covering almost every country in western Europe. Concerned with logistics within Europe and serves industry and trade. See http://www.elalog.org
Electronic Data Interchange	C EDI	Computer-to-computer communication between two or more companies that such companies can use to generate bills of lading, purchasing orders, and invoices. It also enables firms to access the information systems of suppliers, customers, and carriers, and to determine the up-to-the-minute status of inventory, orders, and shipments.
EMC	Tr	Export Management Company. A firm that serves as an export agent, paid by a fee or retainer basis.
EMEA	B	Europe, Middle East, Africa.

Suggest New Words http://www.IDII.com/addword.htm

© 2004 Industrial Data & Information Inc.

Glossary of Supply Chain Terminology

Term	Type	Definition
Emergency Order	W	As differentiated by an order type, an emergency order is processed in an expedited fashion.
Emergency Pull Cord	Conv	Vinyl coated cord that runs along the side of the conveyor that can be pulled at any time to stop the conveyor. Used with an Emergency Stop Switch. (Fs)
Emergency Stop Switch	Conv	Electrical device used to stop the conveyor in an emergency. Used with an emergency pull cord. (Fs)
End-Of-The-Line Terminal	Tr	A local terminal which handles the pick-up and delivery of the customer's freight (as opposed to a consolidation center). Also referred to as a "satellite" or "group" terminal. (Cnf)
Ending Inventory	W	Abbreviated EI. A statement of on-hand inventory levels at the end of a period. See Beginning Inventory and Physical Inventory. (KA)
Enterprise Resource Planning	B	The corporate computer systems with accounting, order management, purchasing, and inventory software. Also termed the "Host System", when one is integrating a WMS or TMS with the ERP solution.
EO	Gov	Executive Order.
EOL	Tr	See End-Of-The-Line-Terminal.
EOQ	Pu	Economic Order Quantity. A method to calculate quantity to purchase based on historical needs. One method of many.
EPC	Std	Electronic Product Code (ePC) is 28 digits in length. Retailers and standard groups are moving to this ePC code. See also GTIN and RSS.

Suggest New Words http://www.IDII.com/addword.htm

© 2004 Industrial Data & Information Inc.

Glossary of Supply Chain Terminology

EPM	C		Enterprise Performance Management. Software to show, plan, and adjust the performance of the entire enterprise. The EPM must connect to the current enterprise software (ERP, SCM, WMS, TMS) to evaluate details and produce performance indicators.
EPS	B		Earnings Per Share.
ER	W		Expected Receipt (EM)
ERM	B		Employee Relationship Management. Software to assist in employee information & management thereof. CRM vendor Siebel (Nasdaq: SEBL) has produced an ERM software module in early 2003.
ERM	B		Enterprise Relationship Management. See also CRM, PRM, and SRM.
ERP	B		See Enterprise Resource Planning.
ERP II	B		Enterprise Resource Planning II. Some view Advanced ERP as ERP. A matter of semantics. Common term is still ERP.
Error Rate	Uom		A percentage of total items picked or shipped in a distribution facility that are not as ordered. The error rate can also be viewed as 100% minus the accuracy level. A company with an accuracy rate of 95% will have an error rate of 5%. (KA)
ESCF	Org		European Supply Chain Forum. See http://www.tm.tue.nl/efgscm/
EST	B		Estimate or Estimated. Common abbreviation used in business and in computer programming.
Estoppel	Tr		The doctrine of equitable estoppel means that he who by his language or conduct leads another to do what he would not otherwise have done, shall not subject such person to loss or injury by disappointing the expectations upon which he acted. (WA)

Suggest New Words http://www.IDII.com/addword.htm

© 2004 Industrial Data & Information Inc.

Glossary of Supply Chain Terminology

ETA	Tr	Estimated Time of Arrival.
ETC	Tr	See Export Trading Company.
ETD	Tr	Estimated Time of Departure.
Ethernet	C	Ethernet is a very common LAN access method. The network administrator may configure the Ethernet LAN to contain many types of computer hardware on it, including various computers, printers, access points, hubs, routers, and more. The computers may be running the same or different operating systems and all be connected on the same Ethernet LAN. This permits one PC user to access multiple computers & devices, if his security levels permit such. . See IEEE, 802.11, 802.11a.
ETO	Mfg W	Engineer-To-Order. A product that is designed only when an order requires it. These are usually highly engineered goods and non-commodity items. See also CTO and MTS.
EULA	B C	End User License Agreement. A contractual agreement for usage of a software package or operating system.
Exception	Tr	1) A shortage, overage, or damage to a shipment. 2) A notation of such conditions on a freight bill, bill of lading or unloading checksheet. (Cnf)
Excess Stock	B	Any quantity of inventory, either held or on order, which exceeds known or anticipated forward demand to such a degree that disposal action should be considered.
Exchange Pallet	Pa	A pallet intended for use among a designated group of shippers and receivers where ownership of the pallet is transferal with the ownership of the unit load; common pool pallet. (P1)

Suggest New Words http://www.IDII.com/addword.htm

© 2004 Industrial Data & Information Inc.

Glossary of Supply Chain Terminology

Exculpatory Clause	Tr	A contractual provision relieving a party from any liability resulting from a negligent or wrongful act. (WA)
Ex-Dec, X-Dec	W	Export Declaration Documents.
Exempt Carrier	Tr	For-hire motor carrier exempt from ICC economic regulation. (Cnf)
Exempt Circular	Tr	A railroad's tariff applying on commodities exempt from government controls. (WA)
Exempt Positions	B	Positions as defined by the Fair Labor Standards Act that do not require overtime payments since they are executive, supervisory, or administrative. (KA)
Exit Interviews	B	An interview with an employee who is leaving a company. The purpose is to discover the reasons for the individual's departure. (KA)
Expediting	Tr	Moving shipments through regular channels at an accelerated rate. (Cnf)
Expendable Pallet	Pa	A pallet intended for a series of handlings during a single trip from shipper to receiver; it is then disposed. See Shipping Pallet. (P1)
Expiration Date	W	Date after which merchandise can no longer be shipped. In Expiration date controlled inventory, inventory must be discarded after this date.
Explosion Limits	MSDS	The range of concentration of a flammable gas or vapor (% by volume in air) in which explosion can occur upon ignition in a confined area. The boundary-line mixtures of vapor or gas with air, which, if ignited, will just propagate the flame. (HC)

Suggest New Words   http://www.IDII.com/addword.htm

© 2004 Industrial Data & Information Inc.

Glossary of Supply Chain Terminology

Explosive	MSDS	A chemical that causes a sudden, almost instantaneous release of pressure, gas, and heat when subjected to sudden shock, pressure, or high temperature. (HC)
Exponential Smoothing	Pu	A forecasting model that uses weighted sum of all prior observations. Weights decline exponentially with the age of the operations. (KA)
Export Broker	Tr	A firm that charges fees to assists foreign buyers with domestic manufacturers.
Export Declaration	Tr	Document declaring goods to be exported which is required by the government.
Export Letter of Credit	Tr	See Letter of Credit.
Export Trading Company	Tr	A firm that purchases goods from domestic manufacturers & sources and sells them to foreign markets. The ETC may work on commission rather than purchasing the goods.
Extendable Conveyor	Conv	Roller or wheel conveyor that may be lengthened or shortened within limits to suit operating needs. Standard extended lengths are 20 ft., 30 ft., and 40 ft. (Fs)
Extinguishing Media	MSDS	Specifies the fire-fighting agents that should be used to extinguish fires. (HC)
EZLogic	Conv	Electronic Zero Pressure Logic - See Hytrol EZLogic Components Manual.(Fs)

Suggest New Words http://www.IDII.com/addword.htm

© 2004 Industrial Data & Information Inc.

- F -

FA	EDI	Functional Acknowledgment. Notice of receipt of an EDI transaction. See 997.
F/A	Tr	Free Astray.
FAA	Org	Federal Aviation Agency. US government agency responsible for air traffic and air safety. See http://www.faa.gov/
Facility	Mfg W	The physical plant and storage equipment. Permanent storage bins in a warehouse may be considered part of the facility, whereas material handling equipment may not. (KA)
Fair Labor Standards Act	Gov	Abbreviated FLSA. Federal law that governs employment practices such as wages and overtime. (KA)
FAK	Tr	Freight All Kinds. An agreed upon freight rate for all types of products from a facility.
Family Grouping	W	A method of sorting products, with similar characteristics, into families to be stored in the same area. (KA)
FAQ	B	Frequently Asked Questions. Detailed answers to frequently asked questions. Designed to help the customer 24x7 days a week. Many companies on their websites have FAQ pages.
FAR	Gov	Federal Acquisition Regulation.
FAS	Tr	Freight Alongside Ship.
FAS	Tr	See Free Along Side.

Suggest New Words http://www.IDII.com/addword.htm

© 2004 Industrial Data & Information Inc.

Glossary of Supply Chain Terminology

FAST	Tr	Participants in C-TPAT, working with Canada's Partners in Protection (PIP) receives unique identifiers that make them eligible for expedited processing at the US-Canada border. Carriers and Drivers must be pre-approved. www.customs.ustreas.gov See also C-TPAT, CSI, and HSA.
Fastener	Pa	A mechanical devise for joining pallet components such as nails, staples, bolts or screws. (P1)
Fastener Shear Index	Pa	Relative measure of shear resistance of the pallet fastener. (P1)
FAT	B	Factory Acceptance Test. The client company is testing the stand-alone performance & functionality that is built-into the software solution.
FB	Tr	See Freight Bill.
FCC	Org	Federal Communications Commission. See http://www.fcc.gov/
FCL	Tr	Full Container Load.
FCS	W	First Customer Ship
FCST	B	Forecast
FDA	Org	U.S. Food & Drug Administration. See http://www.fda.org
FDI	Org	Food Distributors International. See http://www.fdi.org/
FDDI	C	Fiber Distributed Data Interface
FEDA	Org	Foodservice Equipment Distributors Association. See http://www.feda.com/
Feeder	Conv	A conveyor adapted to control the rate of delivery of packages or objects. (Fs)
Feeder	Tr	In intermodal moves, a pickup/delivery vehicle or ship. (Cnf)
FEFO	W	Inventory allocation method, first expired, first-out.
FEWA	Org	Farmers Equipment Wholesalers Association. See http://www.fewa.org/
FF	Tr	See Freight Forwarder.

Suggest New Words http://www.IDII.com/addword.htm

© 2004 Industrial Data & Information Inc.

Glossary of Supply Chain Terminology

Term	Cat	Definition
F/G	Mfg W	Finished Goods.
FGI	W	Finished Goods Inventory
FIFO	W	Inventory allocation method, first-in, first-out. This method allocates first items stored are the first items utilized.
Filed Tariffs	Tr	A carrier's tariff filed with a government agency pursuant to a statute or regulation. (WA)
Fill Rate	B	A measurement of how well a warehouse is meeting service objectives. It is calculated by dividing the number of orders filled by the total number of orders within a given period. Also Order Ratio. (KA)
FILO	W	Inventory allocation method, first in, last-out.
FIM	B	Fresh Item Management (EM)
Finished Product Inventory	Mfg W	Products available for shipment to customers. AKA Finished Goods. (KA)
FIPS	Gov Std	Federal Information Processing Standards.
Fire Wall	W	A wall made of fire-resistant material to prevent a fire from spreading. (KA)
FIRMS	Tr	Facilities Information and Resource Management System code. Identifies the U.S. Customs Service facility where merchandise is located.
Firmware	C	Memory in a computer that retains information when the power is off. The memory may be read-only (ROM), programmable read-only (PROM), erasable programmable read-only (EPROM), or electrically erasable read-only (EEPROM). (KA)
First In, First Out	W	Inventory allocation method, first-in, first-out. This method allocates first items stored are the first items utilized. Abbreviated FIFO
FISA	Org	Food Industry Suppliers Association. See http://www.iafis.org/

Suggest New Words http://www.IDII.com/addword.htm

© 2004 Industrial Data & Information Inc.

Glossary of Supply Chain Terminology

FISH	W	First In Still Here. A funny, but real expression of very slow movers that needs to be dealt with.
Fishyback	Tr	Transporting motor carrier trailers and containers by ship. (Cnf)
FIT	B Tr	Fabrication in Transit.
Fitness	Gov	An applicant for certain government licenses must prove that it is a law-abiding citizen with good moral and other character standards. (WA)
Fixed Beam Scanner	C	A barcode scanner that is stationary and reads codes as items move past it. Also Fixed-Position Scanner. (KA)
Fixed Position Scanner	C	See Fixed Beam Scanner. (KA)
Flammable	MSDS	A chemical that includes one of the following categories: Liquid, flammable--Any liquid having a flash point below 100 °F (37.8 °C), except any mixture having components with flash points of 100 °F (37.8 °C) or higher, the total of which make up 99 percent or more of the total mixture volume. Solid, flammable--A solid, other than an explosive, that can cause fire through friction, absorption of mixture, spontaneous chemical change, or retained heat from manufacturing or processing, or that can be readily ignited and, when ignited, will continue to burn or be consumed after removal from the source of ignition. (HC)
Flapper Gate	Conv	A hinged or pivoted plate used for selectively directing material handled. (Fs)

Suggest New Words http://www.IDII.com/addword.htm

© 2004 Industrial Data & Information Inc.

Glossary of Supply Chain Terminology

Term	Cat	Definition
Flash Point	MSDS	The temperature at which a liquid will give off enough flammable vapor to ignite. The lower the flash point, the more dangerous the product. A "flammable liquid" is a solution with a flash point below 100 °F (37.8 °C). Flash point values are most important when dealing with hydrocarbon solvents. The flash point of a material may vary depending on the method used, so the test method is indicated when the flash point is given. (HC)
Flatbed	Tr	A trailer without sides used for hauling machinery, steel beams, and other lengthy or bulky items. (WA)
Flatcar	Tr	A railcar without sides, used for hauling machinery, steel, etc. (WA)
Flat Face Pulley	Conv	A pulley on which the face is a straight cylindrical drum, i.e. uncrowned. (Fs)
Floor Supports	Conv	Supporting members with vertical adjustments for leveling the conveyor. (Fs)
Flow	Conv	The direction of travel of the product on the conveyor. (Fs)
Flow Rack	W	In one side and out the other. Product stored in this way is necessarily FIFO within the rack itself.
Flow Through	W	See Cross Dock.
Flow Through Rack	W	In one side and out the other. Product stored in this way is necessarily FIFO within the rack itself.
FLSA	Gov	See Fair Labor Standards Act. (KA)
FLT	W	Fork Lift Truck. See Forklift. (EM)
Flush Pallet	Pa	A pallet with deckboards flush with the stringers, stringer-boards, or blocks along the sides of the pallet. (P1)

Suggest New Words http://www.IDII.com/addword.htm

© 2004 Industrial Data & Information Inc.

Glossary of Supply Chain Terminology

FMCSA	Org Tr	Federal Motor Carrier Safety Administration. US government agency responsible for overseeing motor carrier safety & guidelines. See http://www.fmcsa.dot.gov/
FMS	Tr	Freight Management Systems. See TMS.
FOB	Tr	Free On Board. The customer would pay for all shipping costs from the FOB point to the final destination.
FOB	Tr	Freight On Board.
FOB Destination	Tr	Freight costs paid to the destination point, title transfers at destination. (Cnf)
FOB Factory	Tr	Title to goods and transportation responsibility transfers from seller to factory. (Cnf)
FOB Origin	Tr	Free on board up to delivery to the carrier. The consignor retains liability for merchandise until it is picked up by the carrier. The consignee is considered to have title to the goods beginning at the time the carrier accepts the shipment. Typically the consignee is responsible for selecting the carrier and for all charges relating to transportation. (KA)
FOB Shipping Point	Tr	Title to merchandise passes to the consignee at the point when the goods are delivered to the transportation provider. (KA)
FOB Vessel	Tr	Title/transportation costs transfer after goods are delivered on vessel. All export taxes/costs involved in overseas shipments are assessed to the buyer. (Cnf)
FoIP	C	Fax over Internet Protocol. A cost saving method to add fax transmission to broadband networks (intranet and internet). See also VoIP.

Suggest New Words http://www.IDII.com/addword.htm

© 2004 Industrial Data & Information Inc.

Glossary of Supply Chain Terminology 313

Force Majeure	Tr	Condition in contract that relieves either party from obligations where major unforeseen events prevent compliance with provisions of agreement. (Cnf)
Forecast	Pu	An estimate or prediction of future demand. (KA)
Forecast Error	Pu	The difference between actual and forecasted demand. (KA)
Foreign Sales Agent	Tr	A firm or person that works to represent & sell your goods in a foreign country. AKA Foreign Sales Representative.
Foreign Trade Zones	Tr	Goods subject to duty may be brought into such zones duty-free for transshipment/storage/minor manipulation/sorting. Duty must be paid when/if goods are brought from a zone into any part of the U.S. (Cnf)
Fork Entry	Pa	Opening between decks, beneath the top deck or beneath the stringer notch to admit forks. (P1)
Forklift	W	A vehicle with horizontal lift that moves product & freight within a warehouse or on the dock. AKA Fork-Lift, Hi-Lo or Lift-Truck.
Forks	W	The flat metal appendages mounted on lift trucks to facilitate the movement of merchandise on pallets. They are generally four to six inches wide, and 42-48 inches long. (KA)
Foreseeable Emergency	MSDS	Any potential occurrence such as, but not limited to, equipment failure, rupture of containers, or failure of control equipment, which could result in an uncontrolled release of hazardous chemical into the testing environment. (HC)
Forward Location	W	See Pick Slot.

Suggest New Words   http://www.IDII.com/addword.htm

© 2004 Industrial Data & Information Inc.

Glossary of Supply Chain Terminology

Term	Type	Definition
Forwarding Agent	Tr	A firm specializing in shipping goods abroad. Payments made for insurance and other expenses are charged to the foreign buyer. (Cnf)
Four-Way Block Pallet	Pa	A pallet with openings at both pallet ends and along pallet sides sufficient to admit hand-pallet jacks; full four-way entry pallet. (P1)
FPM	Uom	Feet Per Minute.
FPO	W	Firm Planned Orders
Frame	Conv	The structure that supports the machinery components of a conveyor. (Fs)
Frame Spacer	Conv	Cross members to maintain frame rail spacing. AKA as Bed Spacer. (Fs)
Free Alongside	Tr	Selling term in international trade. Selling party quotes price including delivery of goods alongside overseas vessel at exporting port. Abbreviated FAS. (Cnf)
Free-Astry	Tr	A shipment that is mis-routed or unloaded at the wrong terminal and is billed and forwarded to the correct terminal free of charge. (Cnf)
Free On Board	Tr	Loaded aboard carrier's vehicle at point where responsibility for risk/expense passes from seller to buyer. Abbreviated FOB (Cnf)
Free Span	Pa	The distance between supports in a warehouse rack. (P1)
Free-Standing Rack Structure	W	A storage rack supported only by the floor. Free standing racks are not attached to the ceiling, walls, or any other part of the warehouse. (KA)
Free Time	Tr	The period freight will be held before storage charges are applied. The period allowed for the owner to accept delivery before storage charges begin to accrue. (Cnf)

Suggest New Words   http://www.IDII.com/addword.htm

© 2004 Industrial Data & Information Inc.

Glossary of Supply Chain Terminology

Freezer	W	Temperature-controlled (below 32°) storage space for perishable items. Also Cooler. (KA)
Freight	Tr	In North America, freight refers to shipments over 150 lb. to be shipped via LTL, TL, Ocean, Rail, or Air freight. Freight is in contrast to parcel shipments, for small packages, under 150 lb.
Freight Audit	Tr	See Auditing.
Freight Bill	Tr	Carrier's invoice for payment of transportation charges. Abbreviated FB. See also Bill Of Lading.
Freight Broker (Property)	Tr	A person or company responsible for arranging the transportation of goods between points in interstate commerce by motor carrier. As compensation for arranging transport, the broker receives a commission. (KA)
Freight Consolidation	Tr	The merging of shipments from several manufacturers for transport to the same destination under a single bill of lading. This method of shipping provides freight savings for all parties involved. (KA)
Freight Forwarder	Tr	A firm or person that assists with export and other documentation. May assist in the packing and shipping of these goods. Abbreviated FF.
Freight Prepaid	Tr	Notation marked on the face of the bill of lading indicating that the freight bill has been prepaid by the shipper, or that the carrier has extended credit to the shipper. (WA)

Suggest New Words http://www.IDII.com/addword.htm

© 2004 Industrial Data & Information Inc.

FRM	W	Floor Ready Merchandise. Prepared products are received as "floor ready" at the retail store. Before shipping these goods, the warehouse or supplier will add prices, price stickers, tags, security devices, special packaging, etc. so that goods may rapidly be cross docked through RDCs or received directly at the retail stores.
FRU	B	Field Replaceable Unit
FS	B	Field Service
FSL	W	Forward Stock Location. Primary location for picking of a product. AKA as a "pick face".
FSS	Gov	Federal Supply Schedule.
FTD	B	Foreign Trade Division.
FTE	B	Full Time Employees.
FTE	B	Full Time Equivalent.
FTC	Tr	Freight To Customers.
FTL	Tr	Full Truck Load or Full Trailer Load.
FTP	C	File Transfer Protocol. A way to exchange files between computers.
FTSR	B	Foreign Trade Statistical Regulations
FTZ	B Tr	See Foreign Trade Zone.
Functional Acknowledgement	EDI	Notice of receipt of an EDI transaction. Abbreviated FA. See 997.
FVP	Tr	Full value protection. Shippers can opt to pay an additional charge for full replacement value of lost or damaged goods from the carrier.

Glossary of Supply Chain Terminology

- G -

- G -		- G -
GAAP	B	Generally Accepted Accounting Principles.
Gantry Crane	W	Used in warehouse facilities where oversized or very heavy items such as pipe, steel, and heavy machinery must be loaded onto trucks or rail cars. (KA)
GAO	Org	General Accounting Office. See http://www.gao.gov/
Gate	Conv	A section of conveyor equipped with a hinge mechanism to provide an opening for a walkway, etc., manual or spring-loaded. (Fs)
Gateway	Tr	The point at which freight is interchanged/interlined between carriers or at which a carrier joins two operating authorities provision of through service. (Cnf)
GATT	Tr	General Agreement on Tariffs & Trade.
Gaylord	W	Large corrugated carton that is sized to fit a pallet's length and width. Typically used for loose parts or other items stored in bulk. Most commonly found in manufacturing.
GB	C	Gigabyte
GCI	Org	Global Commerce Initiative. Started October 1999. A voluntary global user group to facilitate best practice process recommendations, especially in e-commerce and supply chain. See http://www.globalcommerceinitiative.org
Gear Motor	Conv	A unit which creates mechanical energy from electrical energy and which transmits mechanical energy through the gearbox at a reduced speed. (Fs)
GEIS	Org EDI	General Electric Information Services. Provides EDI VAN service. Now known as GE GXS. See http://www.geis.com

Suggest New Words http://www.IDII.com/addword.htm

© 2004 Industrial Data & Information Inc.

Glossary of Supply Chain Terminology

GFM	Gov	Government Furnished Material. See also GOCO.
Girth	W	The measurement around the sides of the container. (KA)
GLN	Std	Global Location Number. A GLN identifies the entities of the Org. A GLN can identify corporate locations (headquarters, divisions, departments, warehouses, shipping locations, buyers' offices, bill-to offices...). See EAN-UCC for more information.
GMA	Org	Grocery Manufacturers of America. The GMA advances the interests of the food, beverage and consumer products industry on key issues that affect the brand manufacturers. See http://www.gmabrands.org
GMDC	Org	General Merchandise Distributors Corporation. See http://www.gmdc.com/
GMP	Mfg	Good Manufacturing Practice. Good Manufacturing Practices (GMPs) are US, Canada, and European Government regulations that describe the methods, equipment, facilities, and controls required for producing • Human and veterinary products • Medical devices • Processed food
GMPs	Mfg	Good Manufacturing Practice regulations. See GMP.
GNDA	Org	Greater North Dakota Association / Wholesalers Division. See http://www.gnda.com/
GOCO	Gov	Government Owned / Contractor Operated. See also GFM.
Godown	W	A waterfront storehouse in the Orient. (KA)
GOH	W	Goods on hanger. Some wearing apparel is shipped on hangers without packaging. (KA)

Suggest New Words http://www.IDII.com/addword.htm

© 2004 Industrial Data & Information Inc.

Glossary of Supply Chain Terminology

Term	Cat	Definition
GOWI	B	Get On With It. GOWI is an acronym used in messaging other online users.
GPS	Tr	Global Positioning System. Provides accurate latitude and longitude of a position by a series of satellites and receivers to compute the position.
Grandfather Clause	Tr	A provision that enabled motor carriers engaged in lawful operations before the passage of a statute that regulated an industry for the first time, granting applicants a license to continue operations without proving a public need for those services. (referred to as " public convenience and necessity") (WA)
Gravity Bracket	Conv	Brackets designed to permit gravity conveyors to be attached to the ends of a powered conveyor. (Fs)
Gravity Chute	Conv	A chute or trough used to convey commodities by gravity. (KA)
Gravity Conveyor	Conv	Roller or wheel conveyor over which objects are advanced manually by gravity. (Fs)
Green Screen	C	A character based terminal that does not support a graphical user interface (GUI). The early character based terminals had dark green background with white letters – therefore the term "green screen". Character based terminals were very common in 1960's through the 1980's. Know commonly known as a "dumb terminal".
Grocery Pallet Council	Org	A group that provided a standard set of regulations for pallet size and specifications. (KA)
Gross Weight	Tr	Weight of the shipment including packing material. See also Net Weight.
Gross Weight	W	Weight of inventory item with packing and container. See also Gross Weight.
GSA	Org	General Services Administration. See http://www.gsa.gov/

Suggest New Words http://www.IDII.com/addword.htm

© 2004 Industrial Data & Information Inc.

Glossary of Supply Chain Terminology

GT	Uom	Gross Ton.
GTIN	Std	Global Trade Item Number. A 14 digit code representing a product, which most retailers and manufacturers are using or implementing now. The GTIN can be used to synchronize data via UCCnet's GLOBAL registry. See also UCC, ePC, RSS.
Guard Rail	Conv	Members paralleling the path of a conveyor and limiting the objects or carriers to movement in a defined path. (Fs)
GUI	C	Graphical User Interface.
GWTP	B	Get With The Program. GWTP is acronym used in messaging other online users.
GXS	Org EDI	Global eXchange Service. EDI Van service owned by GE. Previously known as GEIS. See http://www.geis.com

- H -		- H -
Hand-Held Scanner	C	A small, portable scanner that can read the bar code symbol. A wand reader is an example of a hand-held scanner. (KA)
Hand (Wheel) Jack Opening	Pa	Space provided in the bottom deck to allow pallet jack wheels to bear on the floor. (P1)
Handling	Pa	A single pick-up, movement and set-down of a loaded or empty pallet. (P1)
Handling	W	The movement of materials or merchandise within a warehouse. (KA)
Handling Charge	W	A charge for ordinary warehouse handling operations. (KA)
Handling Costs	W	The costs involved in warehouse handling. (KA)
Hand Truck	W	A device used for manually transporting goods. A metal plate is slid under the load, then truck and load are tilted toward the operator and moved. There are two varieties: The western type has its wheels located within the side rails, while the eastern type places the wheels located outside the side rails. (KA)
Hard Allocation	W	Detailed allocation of inventory that specifies the exact bin, license plate, lot, and/or serial numbers. Hard allocation is done by the warehouse management system at wave release or pick release. Soft allocation was done previously, which committed a specified quantity only.
Hard Disk	C	An information storage medium built into computers that can store large amounts of data. Also known as a Disk Drive. (KA)
Hardened-Steel Nail	Pa	Heat-treated and tempered steel pallet nail with a MIBANT angle between 8 and 28 degrees. (P1)

Suggest New Words http://www.IDII.com/addword.htm

© 2004 Industrial Data & Information Inc.

Glossary of Supply Chain Terminology

Hardware	C	A computer term that denotes the machinery that makes up a computer system. It includes video screens, memory units, printers, etc. (KA)
Hardwood	Pa	Wood from broad-leaved species of trees (not necessarily hard or dense). (P1)
Hazard Class	Tr	A code indicating type of hazardous material. Normally utilized for carriers that indicate which of their services allow certain hazard classes. E.G. UPS 1 Day AIR services do NOT allow aerosols or combustible products.
Hazard Ratings	MSDS	Material ratings of one to four, which indicate the severity of hazard with respect to health, flammability, and reactivity. (HC)
Hazard Warnings	MSDS	Any words, pictures, symbols, or combination thereof appearing on a label or other appropriate form of warning which conveys the hazards of the chemical(s) in the container(s). (HC)
Hazardous Material	MSDS	In a broad sense, any substance or mixture of substances having properties capable of producing adverse effects on the health or safety of a human being. (HC)
HazMat	Tr W	Hazardous Material. Product is classified hazardous by a government agency or a carrier. Special handling is required on some hazardous items, which may include segregation, separation wall, temperature control, and/or limitations on how the product may be shipped. See also MSDS and Hazard Class.
HCM	B	Human Capital Management.
HD	C	High Density. Usually refers to the 1.44 Megabyte 3 ½ " floppy disks that are of high density format.

Suggest New Words   http://www.IDII.com/addword.htm

© 2004 Industrial Data & Information Inc.

Glossary of Supply Chain Terminology

HDMA	Org	Healthcare Distribution Management Association. See http://www.healthcaredistribution.org/
Health Hazard	MSDS	A chemical for which there is statistically significant evidence, based on at least one study conducted in accordance with established scientific principles, that acute or chronic health effects may occur in exposed employees. The term "health hazard" includes chemicals which are carcinogens, toxic or highly toxic agents, reproductive toxins, irritants, corrosives, sensitizers, hepatoxins, nephrotoxins, neurotoxins, agents that can act on the hematopoietic system, and agents which damage the lungs, skin, eyes, or mucous membranes. (HC)
Helical Nail	Pa	Helically (continuous spiral) threaded pallet nail. See also Drive Screw Nail. (P1)
HIA	Org	Hobby Industry of America. See http://www.hobby.org/
HIBC	Std	Health Industry Bar Code.
HIDA	Org	Health Industry Distributors Association. See http://www.hida.org/
HIDC	Org	Holland International Distribution Council. See http://www.hidc.nl/
High	W	Number of Layers per Pallet
Hi-Lo	W	Another name for a forklift, or a vehicle with horizontal lift. A vehicle to move freight within a warehouse or on the dock. AKA Fork-Lift, Towmotor, or Lift-Truck. See Fork-Lift.
HMF	Tr	Harbor Maintenance Fees
HMI	Mfg W	Human-Machine Interface
Hog Rings	Conv	Rings used to hold the shaft in a roller. (Fs)
Hoist	W	An apparatus for lifting or lowering a load. (KA)

Suggest New Words http://www.IDII.com/addword.htm

© 2004 Industrial Data & Information Inc.

Glossary of Supply Chain Terminology

Honeycomb Factor	W	A term that describes the amount of storage space lost to partially-filled rows or to pallet positions not fully filled. (KA)
Honeycombing	W	A waste of space that results from partial depletion of a lot and the inability to use the remaining space in this area. (KA)
Hopper	Mfg	A funnel-shaped box that is narrower at the bottom than the top so that it will direct material to a conveyor, feeder, or chute. (KA)
Horizontal Carousel	W	See Carousel.
Horizontal Floor Space	Conv	Floor space required for a conveyor. (Fs)
Host System	B	See Enterprise Resource Planning. A "host system" refers to the current enterprise software residing on 1 or more computers. When planning for additional software, one must interface to the host system. When planning for new enterprise wide software, one must convert data from the current host system into the new software.
Hot Load	Tr	A rush shipment. (KA)
House Air Waybill	Tr	An air waybill that an air freight forwarder issues. All house air waybills are covered by a master air waybill which represents all the shipments tendered by an air freight forwarder to an airline. (WA)
Housekeeping	W	Maintaining an orderly environment for preventing errors and contamination in the warehousing process. Good housekeeping includes keeping aisles clear, trash disposal, sweeping, orderly stacking, and removal of damaged goods. (KA)
HQ	B	Headquarters

Suggest New Words http://www.IDII.com/addword.htm

© 2004 Industrial Data & Information Inc.

Glossary of Supply Chain Terminology

HR	B		Human Resources. Either a department within the company dealing with management of employees or a software module that this department heavily utilizes.
HSA	Tr		Homeland Security Act. This US Government Act created the Department of Homeland Security (DHS) and brings the Coast Guard, the Transportation Security Administration (TSA) and US Customs under the DHS. http://thomas.loc.gov. See also C-TPAT, CSI, FAST, and TSA.
HTH	B		Hope This Helps. HTH is an acronym used in messaging other online users.
HTML	C		HyperText Markup Language. A tag language used in building web pages. Web browsers take a web page (ending with .html or .htm), interpret the tags, and display the web page. HTML can be viewed in an editor, whether one uses Notepad, Microsoft Word, or a professional website building tool, such as Macromedia's Dreamweaver.
HTTP	C		HyperText Transfer Protocol. The protocol for moving hypertext files (web pages, etc.) over the Internet.
HTS	Tr		Harmonized Tariff Schedule
HU	W		Handling Unit.
HW	C		Hardware. In computer lingo, this is any electronic device (computer, router, bridge, printer, terminal, screen, scanner...).
Hydrostatic Roller Conveyor	Conv		A section conveyor in which the rollers are weighted with liquid to control their rotation and thereby the speed of movement. (KA)

Suggest New Words http://www.IDII.com/addword.htm

© 2004 Industrial Data & Information Inc.

Glossary of Supply Chain Terminology

- I -		- I -
I2	Org	Software company providing TMS solutions worldwide. Manugistics and i2 Technologies are considered leaders in the TMS software industry. See http://www.i2.com.
I18N	C	See Internationalization.
IAPD	Org	International Association of Plastics Distributors. See http://www.iapd.org/
IARW	Org	International Association of Refrigerated Warehouses. See http://www.iarw.org/
IATA	Org	International Air Transport Association. See http://www..iata.gov/
IAW	B	In Accordance With
IC	C	Integrated Circuit
ICC	Org	Interstate Commerce Commission. An independent regulatory agency that implemented federal economic regulations controlling railroads, motor carriers, pipelines, domestic water carriers, domestic surface freight forwarders, and brokers. The I.C.C. legislated out of existence in 1996 with the passage of the Interstate Commerce Commission Termination Act. (WA)
ICCTA	Gov	Interstate Commerce Commission Termination Act of 1995. (WA)
ICPA	Org Tr	International Compliance Professionals Association (ICPA) is an organization for trade compliance professionals. See http://www.int-comp.org/
ID	B	Identification number or unique identification code value. Common abbreviation used in business and in computer programming.
IDA	Org	Independent Distributors Association. See http://www.idaparts.org/

Suggest New Words http://www.IDII.com/addword.htm

© 2004 Industrial Data & Information Inc.

Glossary of Supply Chain Terminology

IDEAS	W	Internet-based Data Envelopment Analysis System. An online benchmarking tool for warehouse and distribution center to get a "system view" of their warehouse performance. This is a self-assessment tool – developed in 2001 by Georgia Tech and MHIA. IDEAS are located at http://www.isye.gatech.edu/ideas.
IE	Tr	Immediate Export
IEEE	Org	Institute of Electrical and Electronic Engineers. An organization for setting standards for computers and communications. See 802.11, 802.11A. See http://www.ieee.org.
IFB	Gov	Invitation For Bid. See also SIR & SOW.
IFO	W	Income From Operations
IIPMM	Org	Irish Institute of Purchasing Materials Management. See http://www.iipmm.ie/
IL	B	Indirect Labor
ILS	Tr	Integrated Logistics Support.
IMDA	Org	Independent Medical Distributors Association. See http://imda.org/
IMHO	B	In My Humble Opinion. IMHO is an acronym used in messaging other online users.
Inadvertence Clause	Tr	A rule in motor carrier tariffs whereby the carrier declines to accept specified commodities valued over a specific amount. However, the rule states that if the carrier "inadvertently" accepts such an item, the carrier's liability for loss or damage will automatically be limited to the stated dollar amount (such as $5.00 per pound, for instance). (WA)

Suggest New Words http://www.IDII.com/addword.htm

© 2004 Industrial Data & Information Inc.

Glossary of Supply Chain Terminology

Term	Cat	Definition
In Bond	Tr	Storage of goods in custody of government/bonded warehouse or carrier from whom goods can be taken only upon payment of taxes/duties to appropriate government agency. (Cnf) See Bonded Warehouse.
Inbound Logistics	Tr	The portion of logistics operations that involves the movement of materials from source to the production plant or warehouse.(KA)
Inclined Conveyor	Conv	A conveyor transporting up a slope. (Fs)
Incline Conveyor Length	Conv	Determined by the elevation change from infeed to discharge versus the degree of incline. (Fs)
Incompatible	MSDS	Materials that could cause dangerous reactions from direct contact with one another. These types of chemicals should never be stored together. (HC)
Incorporation By Reference	Tr	Contract term referring to the document or set of documents that the contract refers to but is not included within the four corners of the document. When incorporation by reference applies, the item referred to is considered by courts to be a legally binding part of the contract. (WA)
Incoterms	Tr	Maintained by the International Chamber of Commerce (ICC), this codification of terms is used in foreign trade contracts to define which parties incur the costs and at what specific point the costs are incurred. (WA)
Indirect Costs	B	Costs that can not be directly associated with specific goods or services such as utilities, marketing, and staff functions. These are typically allocated to a final product through an overhead account. (KA)
Infeed End	Conv	The end of a conveyor nearest the loading point. (Fs)

Suggest New Words http://www.IDII.com/addword.htm

© 2004 Industrial Data & Information Inc.

Glossary of Supply Chain Terminology

Ingestion	MSDS	The taking in of a substance through the mouth. (HC)
Inhalation	MSDS	The breathing in of a substance in the form of a gas, vapor, fume, mist, or dust. (HC)
Inner Deckboard	Pa	Any deckboard located between the end deckboards. (P1)
Inner Packaging	W	Materials such as paper, foam, or wood shavings used to cushion impacts and prevent the movement of goods within a container. (KA)
Insurance	Tr	Insurance is defined as a contract pursuant to which one party, the insurer, undertakes for compensation, the premium, or the risk of financial loss to another, the insured, against the loss or liability arising from a contingent or unknown event. See Hahn v. Oregon Physicians Serv. et al., 689 F.2d 840, 842 (9th Cir. 1982). (WA)
Inter Alia	Tr	A latin phrase meaning "among other things". (WA)
Interchange	Tr	Passing freight from one carrier to another between lines. (Cnf)
Interleave	W	Multi-tasking. The warehouse worker will be directed to do multiple types of tasks in one trip.
Interleaved 2 of 5	Std	Barcode symbology.
Interline Freight	Tr	Freight moving from origin to destination over two or more transportation lines. (Cnf)
Intermediary	Tr	The term "intermediary" refers to third parties that play a role in arranging for, or providing transportation by carriers for shippers. Intermediaries take the form of truck brokers, foreign, surface, and air freight forwarders, intermodal companies, and third party logistics providers. (WA)

Suggest New Words http://www.IDII.com/addword.htm

© 2004 Industrial Data & Information Inc.

Term	Type	Definition
Intermediate Bed	Conv	A middle section of conveyor not containing the drive or tail assemblies. (Fs)
Intermodal	Org	Intermodal Association of North America. A transportation method that combines several MODES of transportation in combination to ship something to its destination. i.e. TL, Ocean, TL. See http://www.intermodal.org/
Intermodal Transportation	Tr	Using more than one mode to deliver shipments. For example, rail or ocean vessel carriage of tractor-trailer containers. (Cnf)
Internal Costs	W	Those costs generated within the facility and directly under the control of warehouse management these include storage, handling, clerical services, and administration. (KA)
Internationalization	C	Internationalization is the process of building software so it is not based on the assumptions of one language or country. The internationalized software product can display and process locale-dependent information such as date, time, address, and number formats properly to he user. See also Localization.
Interpolate	Conv	To compute intermediate values. (Fs)
Interstate	Tr	The transportation of persons or property between states; in the course of the movement, the shipment crosses a sate boundary. (WA)
Interstate Commerce	B	Exchanging goods between buyers and sellers in two or more states. (Cnf)
Intrastate Commerce	B	When all business between buyers and sellers is carried on within state. (Cnf)
INV	B	Inventory

Suggest New Words http://www.IDII.com/addword.htm

© 2004 Industrial Data & Information Inc.

Glossary of Supply Chain Terminology

Inventory Control	B	The activities and techniques associated with maintaining the optimal level and location of raw materials, work-in-progress, and finished goods in a supply chain. (KA)
Invoice	B	A sales document evidencing the points of purchase, such as quantity, price, weight, etc. (WA)
I/O	C	Input / Output.
IOLT	Org	Institute of Logistics and Transport. See http://www.iolt.org.uk/
IP	B	Intellectual Property (EM)
IP	C	Internet Protocol.
IP Address	C	An unique number for a computer or computer device on a TCP/IP network. Files & data messages are routed based on the destination IP address. Compare this to mail service or parcel delivery, where the letter/shipment, is routed until it reaches its correct destination. A sample IP address is 66.221.212.103 that reaches the computer for IDII.com.
IPO	B	Initial Public Offering.
IPS	Uom	Inches Per Second
IRR	B	Internal Rate of Return
Irritant	MSDS	A substance which, by contact in sufficient concentration for a sufficient period of time, will cause an inflammatory response or reaction of the eye, skin, or respiratory system. The contact may be a single exposure or multiple exposure. (HC)
IS	Org	Information Systems. See also IT, MIS.
ISD	Org	Independent Sealing Distributors. See http://www.isd.org/
ISDN	C	Integrated Services Digital Network.

Suggest New Words http://www.IDII.com/addword.htm

© 2004 Industrial Data & Information Inc.

Glossary of Supply Chain Terminology

ISO	Org	International Standards Organization. A worldwide federation of national standards bodies from more than 140 countries. ISO was established in 1947. See http://www.iso.ch
ISO 9000	Std	International Standards Organization Standards for Quality Systems.
ISP	C	Internet Service Provider. A company providing access & bandwidth to the internet. ISP's regularly provide login accounts, e-mail accounts, and host websites.
ISSA	Org	International Sanitary Supply Association. See http://www.issa.com/
ISSN	Std	Barcode symbology used to encode ISSN numbers for periodicals outside of North America.
ISTA	Org	International Safe Transit Association. See http://www.ista.org/
ISV	C	Independent Software Vendor. A VAR, Value Added Reseller of computer software and/or computer hardware.
IT	C	Information Technology. A generic term for all types of computer software, hardware, network, peoples that work with computers, and the impact of such.
IT	Tr	Immediate Transportation
ITA	Org	Industrial Truck Association. See http://www.indtrk.org/
ITL	C Tr	International Trade Logistics. Software for managing international shipment & documentation. Some of the leading ITL software solutions are Vastera, G-Log.
ITO	B	Information Technology Outsourcing.
ITP	C	Information Technology Providers (EM)

Suggest New Words http://www.IDII.com/addword.htm

© 2004 Industrial Data & Information Inc.

Glossary of Supply Chain Terminology

ITS	Tr	Intelligent Transportation System. ITS systems are in use throughout the United States to promote drive-through toll collection and highway pre-clearance for truckers. Similar technology is in use at the border between NAFTA countries to promote fast and efficient customs checks. In the USA, the FCC's 1999 decision to create a frequency solely for ITS at the 5.9 GHz spectrum provided a new approach the market could endorse, along with allowing for Internet access (data push-pull at ITS access points). The standard is based on IEEE standard 802.11a, which is currently used in private networks. For Road Access & ITS, the standard's group for ITS slightly altered the standard, calling it 802.11/RA (for Road Access.) See also DSRC and ASTM.
IWFA	Org	International Wholesale Furniture Association. See http://www.iwfa.net/
IWLA	Org	International Warehouse Logistics Association. A for-profit organization assisting 3PL's with warehousing operations. See http://www.iwla.org. Most members of the IWLA are based in North America.

Suggest New Words http://www.IDII.com/addword.htm

© 2004 Industrial Data & Information Inc.

- J -		- J -
J2ME	C	Java 2 Platform, Micro Edition. A development platform that allows programmers to build applications using the Java programming language and related tools. See also JDBC.
JAN	Std	Barcode symbology used in Japan, which is similar to EAN.
JCL	C	Job Control Language
JDBC	C	Java Database Connectivity. Connecting Java programs with data files of any type, normally through JDBC API's.
JIT	Mfg	Just In Time. Methodology for reduced inventories, with mindset of zero waste & defects, to arrive to customer at the right time.
Joint	Pa	Intersection and connection of components, often identified by location within the pallet as the end joint, center joint and corner joint. (P1)
Joint and Several Liability	Tr	Liability that may be apportioned either among two or more parties or to only one or a few select members of the group, at the adversary's discretion; thus each liable party is not individually responsible for the entire obligation. (WA)
Joint Rate	Tr	Agreed upon by two or more carriers, published in a single tariff, and applying between point on line of one and point on line of another. May include one or more intermediate carriers in route. (Cnf)

- K -

Kaizen	Mfg	Japanese term that means continuous improvement. (KA)
Kanban	Mfg	Kanban is a Japanese word for "Just In Time" (JIT) Manufacturing. A Kanban is a signboard or placard used to control the Manufacturing production line.
KDF Cartons	W	Knocked down (flat) carton – unassembled packages. (KA)
Keg	Uom	A small barrel. (KA)
Kickback	Tr	Rebate given by a transportation firm. (KA)
Kilobyte	Uom	The amount of computer memory needed to store approximately one thousand (1,024) characters. See Megabyte. (KA)
Kilogram	Uom	The metric measure of weight equal to 2.2046 pounds. (KA)
Kit	W	A number of separate Stock Keeping Units that are supplied or used as one item under its own Item Number.
Kitting	W	Light assembly of component parts, often performed in a warehouse. (KA)
Knee Braces	Conv	A structural brace at an angular position to another structural component for providing vertical support. (Fs)
Knock-Down	W	Abbreviated KD. When articles are taken apart for the purpose of reducing the cubic space of the shipment, the disassembly process is referred to as a knock-down shipment. (KA)
Knur Thumb Adj. Nut	Conv	A nut used on accumulating conveyors to adjust the pressure required to drive the product, may be turned with-out the use of tools. (Fs)
KPI	B	Key Performance Indicator. Metrics.
KSF	B	Key Success Factor.

Suggest New Words http://www.IDII.com/addword.htm

© 2004 Industrial Data & Information Inc.

- L -

L10N	C	See Localization.
LAA	Org	Logistics Association of Australia. See http://www.logassoc.asn.au/
Lacing	Conv	Means used to attach the ends of a belt segment together. (Fs)
Ladder	Std	A barcode printed vertically so that the bars appear to be a ladder.
Lagged Pulley	Conv	A pulley having the surface of its face crowned with a material to provide for greater friction with the belt. (Fs)
Lagging	Conv	A material applied to the outer surface of pulleys or rollers. (Fs)
LAN	C	Local Area Network. See also Network and WAN.
Landed Cost	B	Total expense of receiving goods at place of retail sale, including retail purchase price and transportation charges. (Cnf)
Laser Scanner	C	Bar-code readers that range in size from hand-held units to larger, fixed-beam scanners. Laser scanners are omnidirectional in that the information contained in the barcode an be read regardless of the orientation of the barcode when scanned. (KA)
Last In, First Out	W	Inventory allocation method called Last In, First Out. Abbreviated LIFO
Latent Defects	Tr	Faults which are not readily apparent through normal diligence. The carrier is not responsible for latent defects. (KA)
Lateral Collapse	Pa	The failure of pallet joints due to extreme forces. The force occurs in a direction perpendicular to the stringerboard. (KA)

Suggest New Words http://www.IDII.com/addword.htm

© 2004 Industrial Data & Information Inc.

Glossary of Supply Chain Terminology

Layer	W	A group of Commodities residing in one horizontal dimension on a Pallet. Each Pallet consists of one or more Layers. During picking, Commodities are generally depleted from one Layer completely, before the next layer is taken.
Lay Time	Tr	Downtime during which a ship is being loaded or unloaded and for which there is no demurrage charge. (KA)
Layout	W	The design of the storage areas and aisles of a warehouse. (KA)
Lead Time	B	The period of time that elapses between the time an order is placed and the time it is received in storage. This is often called the replenishment time. (KA)
Lead Time Inventories	B	A quantity that is sufficient to meet demand between the time a replenishment is ordered until it is delivered. See Reorder Point Quantity. (KA)
Learning Curve	B	A term used to describe the ability of individuals to become better at their job and more productive as they gain experience. (KA)
Legal Weight	Tr	The weight of the goods and the interior packaging but not the container. (KA)
Length	Pa	Refers to the stringer or stringerboard (in block pallets) length; also refers to the first dimension given to describe a pallet, i. (P1)e. (P1), 48" x 40", where 48" is the pallet stringer/stringerboard length. (P1)
LC	Tr	See Letter of Credit.
LCL	Tr	Less than Container Load. International shipments utilize containers. When the shipment is less than a full container, the freight will be calculated as LCL. See Less Than Container Load.

Suggest New Words http://www.IDII.com/addword.htm

© 2004 Industrial Data & Information Inc.

Glossary of Supply Chain Terminology

Term	Cat	Definition
L&D	Tr	Loss and damage. This term is usually applied when loss or damage is discovered at delivery time. The term "located loss or damage" is used when the damage occurs at an *identifiable* time. (KA)
LD 50	MSDS	See Lethal Dose 50.
LDOR	Tr	Local Delivery Operations for Replenishment (EM)
Lean	Mfg	Short keyword meaning Lean Manufacturing. (ER)
LED	C	Light Emitting Diode
LEL, LFL	MSDS	Lower explosive limit, or lower flammable limit, of a vapor or gas; the lowest concentration (lowest percentage of the substance in air) that will produce a flash of fire when an ignition source (heat, arc, or flame) is present. At concentrations lower than the LEL, the mixture is too "lean" to burn. See UEL.. (HC)
LES	Tr W	Logistics Execution Systems. Involves transportation management (TMS) and warehouse management (WMS) software. AKA Supply Chain Execution (SCE) or Logistics Resource Management (LRM).
LESA	Org	Logistics Execution Systems Association. See MHIA.
Less Than Container Load	Tr	A quantity of freight less than that required for the application of container rate.
Less Than Truck Load	Tr	A quantity of freight less than that required for the application of truckload rate.
Lessee	Tr	Party or company with legal possession/control of vehicle (with/without driver), or other equipment owned by another under terms of lease agreement. (Cnf)

Suggest New Words http://www.IDII.com/addword.htm

© 2004 Industrial Data & Information Inc.

Glossary of Supply Chain Terminology

Term	Type	Definition
Lessor	Tr	Party or company granting legal use of vehicle (with/without driver), or other equipment to another party under terms of lease agreement. (Cnf)
Let Down	W	Replenishment. Moving inventory from reserve storage to the active picking slots below. Normally the reserve storage is above the forward picking slot, therefore this replenishment is called a "let down".
Lethal Dose 50	MSDS	A single dose of a material expected to kill 50 percent of a group of test animals. The dose is expressed as the amount per unit of body weight, the most common expression being milligrams of material per kilogram of body weight (mg/kg of body weight). Usually refers to oral or skin exposure. (HC)
Letter of Credit	Tr	Buyer's bank issues these document that guarantees payment to vendor, with conditions that goods are shipped properly and proper documents are presented. Abbreviated L/C, LC, or LOC.
LF	W	Logistics Fulfillment (EM)
LGMDA	Org	Lawn & Garden Marketing and Distribution Association. See http://www.lgmda.org/
License Plate Number	W	A unique number that is applied to a container to rapidly identify the contents (products) within. The LPN is normally printed as a barcode on a label or the unique id of the RFID tag.
Lien	Tr	Logal device whereby the lien holder is able to secure the owner's goods until payment or resell them. The lien must be properly filed and attached or the lien holder may be found guilty of conversion. (WA)
Life to First Repair	Pa	Output from PDS program which is equivalent to the number of trips the pallet will last before needing repair. (P1)

Suggest New Words   http://www.IDII.com/addword.htm

© 2004 Industrial Data & Information Inc.

Glossary of Supply Chain Terminology

Term	Cat	Definition
LIFO	W	Inventory allocation method called Last In, First Out.
Limit Switch	Conv	Electrical device used to sense product location. (Fs)
Linehaul	Tr	Movement of freight between cities that are usually more than 1,000 miles apart, not including pickup and delivery service. (Cnf)
Lift Truck	W	Another name for a forklift, or a vehicle with horizontal lift. A vehicle to move freight within a warehouse and dock. AKA Fork-Lift, Towmotor, or Hi-Lo. See Fork-Lift.
Line Balancing	Mfg	Readjusting product mix and other resources on a production line to achieve the greatest efficiency and consistency. (KA)
Line Load	Pa	The weight of a unit load concentrated along a narrow area across the full length or width of the pallet. (P1)
Lineside Warehouse	W	A supplier warehouse positioned as close as possible to the production location to facilitate Just In Time manufacture.
Linux	C	A free (or low cost) operating system. It is a derivative of Unix. Linux is available & supported from many sources, including Red Hat Software. Website for Red Hat Software is http://www.redhat.com. Many major vendors are now supporting Linux including IBM, Sun, and others.
Live Rack	W	A storage rack constructed to allow items to move unaided toward the picking point. The rack is slanted so that the picking point is lower than the rear loading point, allowing gravity to draw items to the front. A roller conveyor or other low-friction surface supports the merchandise. (KA)

Suggest New Words http://www.IDII.com/addword.htm

© 2004 Industrial Data & Information Inc.

Glossary of Supply Chain Terminology

Term	Cat	Definition
Live Roller Conveyor	Conv	A series of rollers over which objects are moved by the application of power to all or some of the rollers. The power-transmitting medium is usually belting or chain. (Fs)
LLP	Org	Lead Logistics Provider. A master contractor who manages an entire outsourced logistics network for a company. LLPs traits include multi-modal services, advanced technology, international reach, and able to manage complex requirements. AKA as a 4PL.
LMI	Org	Logistics Management Institute. See http://www.lmi.org/
LN	B	Line. Common abbreviation used in business and in computer programming.
Load Bearing Surface	Pa	Actual area of material in contact with and supporting a unit load. (P1)
Load Factor	Tr	1) The weight in pounds loaded onto a trailer. 2) A term used loosely to describe the "compactness" or over "good usage" of trailer space. (Cnf)
Loading Allowance	Tr	A tariff provision which provides an allowance, usually a fixed sum per hundredweight, to a shipper for loading a carrier's trailer. (Cnf)
LOB	B	Line of Business.
LOC	B	Location. Common abbreviation used in business and in computer programming.
LOC	Tr	See Letter of Credit.

Suggest New Words http://www.IDII.com/addword.htm

© 2004 Industrial Data & Information Inc.

Glossary of Supply Chain Terminology

Localization	C	Localization. Localization is often abbreviated as "l10n "because of the 10 letters between the "l "and the "n ". After a product has been internationalized, it is then localized. Localization is the process of adapting and translating the product into another language to make it linguistically and culturally appropriate for a particular country. Translation is one of the activities during localization.
Location	W	The place in a warehouse where a particular product can be found. (KA)
Location Audit	W	A systematic verification of the location records of an item or group of items by checking the actual locations in a warehouse or storage area. (KA)
Logistics	Tr W	Logistics plans, implements, and controls the efficient, effective forward and reverse flow and storage of goods, services, and related information between the point of origin and the point of consumption in order to meet customers' requirements. (Cnf)
LOI	B Tr	Letter of Intent.
LOL	B	Laughing Out Loud. LOL is an acronym used in messaging other online users.
LO/LO	Tr	Lift On, Lift Off vessel. The method or process in which cargo is loaded unto and unloaded from an ocean vessel.
Longshoreman	Tr	A person who loads and unloads marine vessels. (KA)
Long Ton	Uom	Equivalent to 2,240 pounds or 20 long hundredweights. Also called gross ton. (Cnf)

Suggest New Words   http://www.IDII.com/addword.htm

© 2004 Industrial Data & Information Inc.

Glossary of Supply Chain Terminology

Loose	W	The number of remaining Commodities left on a particular Layer. Ex. Tie, High Loose: 7, 3, 1 means 7 boxes per layer, 3 layers high, and 1 loose box on top of the 3rd layer.
Lost Time Factor	Uom	The complement of the productivity factor. It can also be viewed as 100 percent minus the productivity factor. A company with a productivity factor of 95 percent will have a lost time factor of 5 percent. (KA)
Lot	W	Product that has a life span with an expiration date, Manufacturing date, and/or code date. A lot is a group of an item that has been made using the same ingredients (like a batch), a production run, or some other grouping to be used for identification purposes, unique from other same items produced in a different batch or run. For instance fabric color in the same SKU may not match, lot to lot.
Lot-For-Lot	Mfg	Lot sizing technique that matches planned orders with the net requirement for a period. (KA)
Lot Number	W	Identifying numbers used to keep a separate accounting for a specific lot of merchandise. (KA)
Lot Number Traceability	W	The ability to track an item or group of items by a unique set of numbers based by a vendor or production run. (KA)
Lot Size	Mfg	The amount of a product ordered. It is often equal to EOQ (quantitiy is to be transported in several shipments). Lot size does not include safety stock. (KA)

Suggest New Words   http://www.IDII.com/addword.htm

© 2004 Industrial Data & Information Inc.

Lottable	W	Specific attributes of a Commodity that, taken together, differentiates like Commodities, and allow like Commodity to be tracked as separate Lots throughout the facility. Owners of the Commodity drive what is considered a Lottable. Examples of Characteristics: color, pack size, original production facility, etc.
LOV	B	List of Values (EM)
Low Lift	Pa	See Pallet Jack. (KA)
LPM	C	Lines Per Minute
LPN	W	License Plate Number. A unique number assigned to a container. A container is any type of container that can hold inventory – such as a pallet, carton, and tote. WMS use & assign LPNs to track inventory.
LRM	Tr W	Logistics Resource Management. Software that includes modules with transportation management, warehouse/inventory management, and contracting/rate negotiations. AKA as Logistics Execution System (LES) and Supply Chain Execution (SCE).
LSP	B	Logistics Service Provider.
LT	Pu	See Lead Time.
LTL	Mfg	Lot To Lot.
LTL	Tr	See Less Than Truckload.
Lumping	Tr	Force a truck driver to hire those other than himself to unload his lading. (WA)

Glossary of Supply Chain Terminology

- M -

- M -		- M -
MAC	C	Macintosh Computer
MAC	C	Media Access Control. For those using wireless 802.11, it is best to use an encryption to keep wireless messages secured, such as AES, DES, MAC, or WEP.
Machine Crowned Pulley	Conv	A pulley in which the crown or vertex has been produced by an automatic, usually computer driven, machine. (Fs)
Machine Readable	C	1) Information that can be read directly from a printed format to an electronic form without human intervention, such as in bar-coding. 2) In computer applications, compiled programs. Human-readable object code Is readable by humans only with the help of programming tools. (KA)
MAD	B	Mean Absolute Deviation.
MAD	Org	Michigan Association of Distributors. See http://www.asa.net/
Magnetic Starter	Conv	An electrical device that controls the motor and also provide overload protection to the motor. (Fs)
Magnetic Strip	C	A type of identification tag that uses a strip of magnetic material attached to a container or to the merchandise itself. The strip is encoded with information that can be read by a magnetic scanner. (KA)

Suggest New Words http://www.IDII.com/addword.htm

© 2004 Industrial Data & Information Inc.

Glossary of Supply Chain Terminology

Term	Type	Definition
Maintenance, Repair, and Operating Items and Supplies	B	Abbreviated MRO. All items used maintaining, repairing and operating a facility. Maintenance items are those necessary to keep the facility in optimum operating condition, such as janitorial supplies. Repair items are those necessary to keep the facility and machinery functioning, such as replacement parts. Operating items are those required for the operation such as fuel and office supplies. (KA)
Management by Objectives	B	Abbreviated MBO. A participative method of motivational management. Goals are determined with input from subordinates. (KA)
Manhattan	Org	Manhattan Associates - provider of supply chain execution solutions worldwide. Manhattan and Red Prairie are considered leaders in the WMS software industry. www.manh.com
Manifest	Tr	A control document used to list the contents (individual shipments) during loading and from which the contents are checked during unloading. (Cnf)
Manual Start Switch	Conv	A simple one-direction switch used to turn the conveyor on or off. (Fs)
Manufacturing Resource Planning	Mfg	Abbreviated MRPII. MRP II includes MRP plus shop floor, accounting, distribution management, and also activities within Manufacturing and the enterprise. MRP II developed in the 1980's and 1990's. See ERP.
Manugistics	Org	A software company providing TMS solutions worldwide. Manugistics and i2 Technologies are considered leaders in the TMS software industry. See http://www.manugistics.com.

Glossary of Supply Chain Terminology

Term	Type	Definition
Marine Lien	Tr	A lien on a vessel given to secure the claim of a creditor who provided maritime services to the vessel or who suffered an injury as a result of the vessel's use. (WA)
Market Dominance	Tr	A railroad's control of traffic for which a shipper has no alternative transportation. (WA)
Marking Machine	W	A machine that imprints or embosses a mark on a label, ticket, tape, package, or tag. (KA)
Master/Agent Principle	Tr	General tort principle generally referring to the ability of the agent, in most cases an employee, to make liable the master, employer, for action that the agent undertook. (WA)
Master BOL	Tr	See BOL. A master BOL is a BOL document listing all items to be delivered. Usually, the master BOL is on top of the separate BOLS to form a "package" for the driver or carrier.
Master Carton	W	A single large carton that is used as a uniform shipping carton for smaller packages. It is used primarily for protective purposes, but also simplifies materials handling by reducing the number of pieces handles. See Master Pack. (KA)
Master File	C	In data processing, a computer file of data used by various programs. For example, the customer master file may be used by the order-entry and billing routines. (KA)
Master Pack	W	An established quantity for a manufacturer's product. The master pack is a carton containing a set number of multiple case quantities.

Suggest New Words http://www.IDII.com/addword.htm

© 2004 Industrial Data & Information Inc.

Glossary of Supply Chain Terminology

Master Trip Lease	Tr	An agreement between either a regulated carrier and a private carrier or two regulated carriers. The master trip lease covers multiple loads handled by the two parties for a period of 30 days or more. The master trip lease outlines the responsibilities of each party regarding licensing, insurance, fuel permits, DOT safety compliance, maintenance, vehicle and driver identification and rates of pay. (KA)
Mat	Pa W	A panel of wood, rubber, or other material that is placed on top of unit loads to allow tight strapping of the load without product damage. (KA)
Material Handling	W	The movement of materials going to, through, and from warehousing, storage, service facility, and shipping areas. Materials can be finished goods, semi-finished goods, components, scrap, WIP, or raw stock for Manufacturing.
Materials Management	W	The functions that define the complete cycle of material flow, from the purchase to the distribution of the finished product. This includes functions of planning production materials, control of work in process, warehousing, shipping, value added services at the warehouse, and distribution.
Material Requirements Planning	Mfg	Abbreviated MRP. MRP uses bills of material, inventory data, and MPS to calculate time-phased materials requirements and recommend release or reschedule of orders for materials. Developed in the 1970's. See MRP II.

Suggest New Words http://www.IDII.com/addword.htm

© 2004 Industrial Data & Information Inc.

Glossary of Supply Chain Terminology

Material Safety Data Sheet	MSDS W	A Material Safety Data Sheet (MSDS) is prepared by a chemical manufacturer, and summarizes available information on the health, safety, fire, and environmental hazards of a chemical product. It also gives advice on the safe use, storage, transportation, and disposal of that product. Other useful information such as physical properties, government regulations affecting the product, and emergency telephone numbers are provided in the MSDS as well. There is a detailed description of how to read an MSDS and a useful glossary of MSDS terms in Hach's Website at http://www.hach.com. (HC)	
Materials Warehouse	W	A warehouse used exclusively for the storage of raw materials. (KA)	
Matsui Amendment	Gov Tr	49 U.S.C. § 10502 (c)(1) requiring railroads to maintain full value rates on exempt traffic if they offer reduced rates. (WA)	
MAX	B	Maximum. Common abbreviation used in business and in computer programming.	
Maxicode	Std	2D barcode symbology used by the United Parcel Service (UPS).	
MB	Uom	Megabyte.	
MBO	B	Management Buy Out.	
MBO	B	Management By Objectives.	
Mbps	Uom	Megabits per second. Commonly used to measure data transfer speeds. Since there are 8 bits per byte, a data transfer speed of 80 Mbps is 10 MB per second.	
MCS	W	Material Control Software. AKA Material Handling Control Software, Warehouse Control Systems, or Conveyor Control Software. This is the software that interfaces the machinery to the WMS or MES.	

Suggest New Words http://www.IDII.com/addword.htm

© 2004 Industrial Data & Information Inc.

Term	Cat	Definition
MDA	Org	Midwest Distributors Association. See http://www.asaonline.org/
MDA	Org	Music Distributors Association. See http://www.musicdistributors.org/
Meet	Tr	A process by which two drivers going in opposite directions, exchange loads in order to keep the freight moving while allowing the drivers to return to their home locations. (KA)
Meltdown	Tr	The failure of an entire system to handle functions like informatics, routing, etc. (WA)
Melting Point	MSDS	The temperature at which a solid substance changes to a liquid state. For mixtures, the melting range may be given. (HC)
Merchandise	B	Goods bought and sold in trade, with the exception of real estate, cash, and negotiable paper. (KA)
Merger	Tr	The combination of two or more carriers into one company that will own, manage, and operate the properties that previously operated separately. (WA)
MES	Mfg	Manufacturing Execution System.
Mesh	W	A panel or fabric with openings (pores) of uniform size. It is designed by citing the number of openings per square inch. (KA)
Metadata	C	Metadata is "data about data". Metadata describes the structure and format that the data is in.
Mfg	Mfg	Manufacturing.
MFLOPS	C	Millions of Floating Point Operations Per Second
MHE	W	Material Handling Equipment.
MHE	W	Mechanical Handling Equipment.

Suggest New Words http://www.IDII.com/addword.htm

© 2004 Industrial Data & Information Inc.

Glossary of Supply Chain Terminology

MHIA	Org	Material Handling Industry of America. See http://www.mhia.org. Within the membership of MHIA, there are divisions. Two of these units are the Logistics Execution Systems Association (LESA) and Order Selection, Staging & Storage Council (OSSSC).
MHZ	C	Mega-Hertz.
MIB	Tr	Master-In-Bond. Paperless Master-In-Bond reporting by carriers that are MIB-approved.
MIBANT	Pa	Morgan Impact Bend Angle Nail Tester.
MIBANT Angle	Pa	The bend angle in a fastener shank when subjected to a MIBANT test. (P1)
MIBANT Test	Pa	Standard impact nail tester used in the pallet and lumber industry as an indication of impact bend resistance of nails or staples. (P1)
MICR	B C	Magnetic Ink Character Recognition. Special ink with magnetic abilities to be read by high peed readers. Commonly used in personal & business checks.
Microcomputer	C	A computer that uses a single-chip microprocessor. Microcomputers and mainframe computers have more power, but microcomputers have evolved so that many are nearly as powerful as mini computers. Personal computers are based on microcomputer architecture. See Personal Computer. (KA)
Middleware	C	Middleware software helps separate application software to talk to each other. I.E., Mercator, IBM's MQ series.
Milk Run	Tr	A pick-up route with multiple stops. (KA)
Millisecond	Uom	One-thousandth of a second. (KA)
MIN	B	Minimum. Common abbreviation used in business and in computer programming.

Suggest New Words http://www.IDII.com/addword.htm

© 2004 Industrial Data & Information Inc.

Glossary of Supply Chain Terminology

Term	Cat	Definition
Mincomputer	C	A general-purpose computer smaller than a mainframe, but capable of serving a number of users. (KA)
Minimum Pressure Accumulating Conveyor	Conv	A type of conveyor designed to minimize build-up of pressure between adjacent packages or cartons. (I38-ACC - 19O-ACC) (Fs)
Min-Max System	W	An order-point replenishment system. The minimum point is the order point and the maximum is the "order-up-to" level. The order quantity is variable to take advantage of usage patterns and lot-sized ordering economies. (KA)
MIPS	C	Millions of Instructions Per Second
MIS	C Org	Management Information Services, Management Information System. The department or division responsible for computer applications, programmers, developers, quality assurance, documentation of applications, and hot-line support staff. AKA IS and IT.
MIT	Org	Massachusetts Institute of Technology. See http://web.mit.edu/
Mix	Tr	Refers to the combination of light, medium and heavy density freight. (Cnf)
Mixture	MSDS	Any combination of two or more chemicals if the combination is not, in whole or in part, the result of a chemical reaction. (HC)
MLP	W	Master License Plate (EM)
Mode	Tr	A method of transporting materials such as truck, rail, air, ocean barge, and intermodal. (KA)

Suggest New Words http://www.IDII.com/addword.htm

© 2004 Industrial Data & Information Inc.

Glossary of Supply Chain Terminology

Modem	C	A computer attachment that allows the transmission of information over telephone lines between computers. A modem converts digital signals to analog for transmission and then converts the signals back to digital at the other end of the phone line. Modem is an acronym for **mo**dulator/**dem**odulator. (KA)
Modified Plessey Code	Std	Barcode symbology.
Mods	C	Abbreviation for Modifications. Computer people frequently use the term "mods" to indicate a minor or major modification is needed. It can used to discuss a specific modification being designed to do specific business functionality, or it can be used to just indicate that the program would need to be changed in order to perform this.
MOQ	Pu	Minimum Order Quantity required by the supplier.
MOS	Tr	Method of Shipment (EM)
MOT	Tr	Method of Transportation.
Motor	Conv	A machine that transforms electric energy into mechanical energy. Standard motors are dual voltage and operate at 1725 RPM. (Fs)
Motor Carrier Act of 1935	Tr	Act of US Congress bringing motor common and contract carriers under ICC jurisdiction. (Cnf)
Motor Carrier Act of 1980	Tr	Act of Congress that deregulated for-hire-trucking. (Cnf)
MP	C	Multi-Processor.
Mpg	Uom	Miles per gallon.
Mph	Uom	Miles per hour.
MPS	C Mfg	Master Production Scheduling.
MPS	Mfg	Master Plan Schedule.
MRO	B	See Maintenance, Repair, and Operating Items and Supplies. (KA)

Suggest New Words   http://www.IDII.com/addword.htm

© 2004 Industrial Data & Information Inc.

Glossary of Supply Chain Terminology

MRO	W	Maintenance, Repair, and Operating supplies.
MRO	W	Maintenance, Repair, and Operations.
MRP	Mfg	Material Requirements Planning. MRP uses bills of material, inventory data, and MPS to calculate time-phased materials requirements and recommend release or reschedule of orders for materials. Developed in the 1970's. See MRP II.
MRP II	Mfg	Manufacturing Resource Planning. MRP II includes MRP plus shop floor, accounting, distribution management, and also activities within Manufacturing and the enterprise. MRP II developed in the 1980's and 1990's. See ERP.
MS	C Uom	Millisecond (one one-millionth of a second).
MS	Mfg	Material Supply
MSDS	MSDS W	See Material Safety Data Sheet.
MSG	B	Message. Common abbreviation used in business and in computer programming.
MSI Plessey	Std	Barcode symbology.
MTBF	Mfg	Mean Time Between Failure. Average running time between failures.
MTD	B	Month To Date. Showing a summary, sub-total, or total value that is a month to date value.
MTF	Mfg	Make To Forecast.
MTO	Mfg W	Make To Order. A product that is manufactured only when an order is confirmed for it. These are usually low volume or highly engineered goods. See also CTO, ETO, and MTS.
MTS	Mfg W	Make To Stock. Products are manufactured and placed into warehouse as finished goods. The quantity built is based upon safety stock and re-order point calculations. See also CTO, ETO, and MTO.

Suggest New Words http://www.IDII.com/addword.htm

© 2004 Industrial Data & Information Inc.

Glossary of Supply Chain Terminology

MTTR	Mfg	Mean Time To Repair. Average time to perform the repair.
Mule	Tr	A small tractor used to move two-axle dollies. The mule is also called a yard tractor. (KA)
Mullen Test	W	A device to test, or the process of testing packaging material to establish its strength. (KA)
Multimodal	Tr	See Intermodal Transportation.
Multi-Tine Fork	W	Attachment to a forklift truck that allows the movement of two pallets side-by-side, rather than one pallet at a time. (KA)
Mutagen	MSDS	Those chemicals or physical effects that can alter genetic material in an organism and result in physical or functional changes in all subsequent generations. (HC)

Suggest New Words http://www.IDII.com/addword.htm

© 2004 Industrial Data & Information Inc.

Glossary of Supply Chain Terminology

- N -		- N -
NACD	Org	National Association of Container Distributors. See http://www.nacd.net/
NACFAM	Org	National Coalition for Advanced Manufacturing. See http://www.nacfam.org/
NAED	Org	National Distribution of Electrical Distributors. See http://www.naed.org/
NAFTA	B	North American Free Trade Agreement. NAFTA is a free trade agreement between Canada, the United States, and Mexico to eliminate trade related tariffs and duties, as well facilitate increased environmental, transportation, and labor-related safety. (WA)
NAGASA	Org	National American Graphic Arts Suppliers Association. See http://www.nagasa.org/
NAHAD	Org	National Association of Hose & Accessories Distributors. See http://www.nahad.org/
NAHSA	Org	North American Horticultural Suppliers Association. See http://www.nahsa.org/
Nail	Pa	Fastener made from endless wire by cutting a point and forming a head at the shank end opposite the point. (P1)
Nail Diameter	Pa	The average diameter of a nail shank. (KA)
Nail Head Diameter	Pa	The average diameter of a nail head. (KA)
Nail Length	Pa	The distance measured parallel to the shank from the top of the head to the point of the nail. (KA)
Nail Shank	Pa	The length of the nail, not including the tip or head. (KA)
NAPM	Org	National Association of Purchasing Management

Suggest New Words http://www.IDII.com/addword.htm

© 2004 Industrial Data & Information Inc.

Glossary of Supply Chain Terminology

NAPSA	Org	National Appliance Parts Suppliers Association. See http://www.napsaweb.org/
NARM	Org	National Association of Recording Merchandisers. See http://www.narm.com/
NARS	Org	National Association of Rail Shippers. See http://www.railshippers.com/
NASFT	Org	National Association for the Specialty Food Trade, Inc. See http://www.nasft.org/
NASGW	Org	National Association of Sporting Goods Wholesalers. See http://www.nasgw.org/
NATCD	Org	National Association of Tobacco and Confectionary Distributors. See http://www.natcd.com/
NATD	Org	North American Association of Telecommunications Dealers. See http://www.natd.com/
National Motor Freight Classification	Tr	A publication for motor carriers containing rules, descriptions, and ratings on all commodities moving in commerce. The NMFTA maintains these codes. Carriers will base their freight rates on the NMFC codes (unless one utilizes FAK).
NAW	Org	National Association of Wholesaler-Distributors. See http://www.naw.org/
NCBAA	Org	National Customs Brokers / Forwarders Association of America. See http://www.ncbfaa.org/
NCC	Org	National Classification Committee.
NDC	Std	National Drug Code. See also GTIN, RSS, and UCC.
Negative Crowned Pulley	Conv	A pulley with raised areas set equally in from each end. This crown is used on tail pulleys 24 in. OAW and wider, and aids in belt tracking. (Fs)

Suggest New Words http://www.IDII.com/addword.htm

© 2004 Industrial Data & Information Inc.

Negotiable Warehouse Receipt	W	A legal certification that listed goods are held in public warehouse. The certificate can be purchased or sold, thus transferring title to the goods. See Non-negoiable Warehouse Receipt. (KA)
Nested	W	The process of packing articles so that one rests partially or entirely within another, thereby reducing the total cubic displacement. (KA)
Nested Solid	W	Articles nested so that the bottom of one rests on the bottom of the lower one. (KA)
Nesting Factor	W	Nesting on products that "fit" within each other, therefore the first item is full cube, the 2nd, 3rd just use the "factor" cube. E.G., Buckets would have a .20 (20%) nesting factor.
NET, .NET	C	Microsoft's .NET initiative.
Net Lift	Conv	The net vertical distance through which material is moved against gravity by a conveyor. (Fs)
Net Storage Area	W	The amount of warehouse space actually used for storage of merchandise. It usually expressed in square feet or meters and excludes aisles, dock areas, staging areas, and offices. (KA)
Net Weight	Tr	Weight of shipment not including packing materials and container. See also Gross Weight.
Net Weight	W	Weight of inventory item without packing and container. See also Gross Weight.

Suggest New Words http://www.IDII.com/addword.htm

© 2004 Industrial Data & Information Inc.

Network		C	A network of computers connected together. Other computer equipment may be on the network, such as printers, hubs, and switches. A network is a small defined size and setup & administered by a computer geek called a System Administrator or Network Administrator. A network is usually attached to other networks. Sometimes the entire set of networks is called the "network". E.G., the internet is a large network. A popular network method of connectivity is TCP/IP. See also LAN, WAN.
NFFA		Org	National Frozen Foods Association. See http://www.nffa.org/
NFPA		MSDS Org	National Fire Protection Association is an international membership organization which promotes/ improves fire protection and prevention and establishes safeguards against loss of life and property by fire. Best known on the industrial scene for the National Fire Codes (16 volumes of codes, standards, recommended practices and manuals developed and periodically updated by NFPA technical committees). Among these is NFPA 704M, the code for showing hazards of materials as they might be encountered under fire or related emergency conditions, using the familiar diamond-shaped labels or placards with appropriate numbers and symbols. (HC)
NFR		B	Not For Resale. An indicator by the manufacturer that this product is not intended to be resold. This version of the product is either free or heavily discounted, while the same product normally sells higher in its regular price range.

Suggest New Words   http://www.IDII.com/addword.htm

© 2004 Industrial Data & Information Inc.

Glossary of Supply Chain Terminology

NGA	Org	National Grocers Association. See http://www.nationalgrocers.org/
NIACAP	Std	National Information Assurance Certification and Accreditation Process.
NIOSH	Org	National Institute for Occupational Safety and Health. OSHA and NIOSH are partnering together for guidelines and regulations for worker safety based on research & advisory committee. See http://www.cdc.gov/niosh/homepage.html
Nip Point Guard	Conv	A guard placed to eliminate points or areas on the conveyor where injuries might occur. (Fs)
NIS	W	Not In Stock.
NIST	Org	National Institute of Standards Technology. See http://www.nist.gov/
NITL	Org	Institute of National Transport Logistics. See http://www.nitl.ie/
NMFC	Tr	See National Motor Freight Classification.
NMFTA	Org	National Motor Freight Traffic Association, Inc. A non-profit organization whose members are primarily LTL motor carriers. The NMFTA maintains the NMFC codes, assigns the SCAC (Standard Carrier Alpha Code) to motor carriers, and more. See http://www.nmfta.org
NMHC	Org	National Materials Handling Center.
NOIBN	Tr	Not otherwise indexed by name. A rate classification that is similar to FAK but not as broad. It covers all commodities that are not specifically described in the tariff. See Freight of all Kinds. (KA)
Noncontiguous	B	A possession of a sovereign nation although not geographically interconnected. (WA)

Suggest New Words http://www.IDII.com/addword.htm

© 2004 Industrial Data & Information Inc.

Glossary of Supply Chain Terminology

Non-Exempt Positions	B	Employees who are not assigned to supervisory or executive positions. These workers are eligible for overtime. (KA)
Non-Negotiable Warehouse Receipt	W	A legal certification that the goods listed on it are now in the custody of the public warehouse. The certificate cannot be bought or sold. (KA)
Non-Read	C W	Failure of scanner to read a barcode or RFID tag.
Non-Reversible Pallet	Pa	A pallet with bottom deckboard configuration different from top deck. (P1)
Non-Vessel Operation Common Carrier	Tr	Consolidates and issues their own BOL for shipment on a vessel. Does not own nor operate the vessel. Abbreviated NVOOC.
Nose Roller	Conv	A small roller, used on power belt curve conveyors, to reduce the gap at the transfer points. (Fs)
Noseover	Conv	A section of conveyor with transition rollers placed in conveyor to provide transition from incline to horizontal or horizontal to incline. (Fs)
Notch	Pa	Cutout in lower portion of the stringer to allow entry for the fork tine, usually 9" in length, 1-1/2" in depth. (P1)
Notched Stringer	Pa	A stringer with two notches spaced for fork-tine entry, (partial four-way entry). (P1)
Notch Filet	Pa	The curvature at either corner of a pallet notch. (KA)
Notch Height	Pa	The distance between the bottom surface of a pallet stringer and the top of the notch. (KA)
Notice of Arrival	Tr	On arrival of freight at destination, notice is sent promptly to the consignee showing number of packages, description of articles, route, rate, weight, etc. (Cnf)
NPD	B	New Product Development.

Suggest New Words   http://www.IDII.com/addword.htm

© 2004 Industrial Data & Information Inc.

Glossary of Supply Chain Terminology

NPTC	Org	National Private Truck Council. See http://www.nptc.org/
NSN	GovW	National Stock Number. See also SKU.
NSP	Mfg	Non-Stock Production system.
NSPI	Org	National Pool & Spa Institute. See http://www.nspi.com/
NSSEA	Org	National School Supply & Equipment Association. See http://www.nssea.org/
NT	C	See Windows NT.
N-Tier	C	Software program that is distributed among separate computers over a network. N-Tier implies client/server program model. The most common is 3-tier, where the user interface programs are on the user's computer, the business logic is on a central server, and the data is stored in a database on a third computer.
NTP	MSDS	National Toxicology Program. The NTP publishes an Annual Report on Carcinogens which identifies substances that have been studied and found to be carcinogens in animal or human evaluations. (HC)
NTP	Org	National Transportation Program, which is part of the DOE in US Government. See http://www.ntp.doe.gov
NUSA	Org	National Unaffiliated Shippers' Association. See http://www.nusa.net/
NVOCC	Tr	Non-Vessel Operation Common Carrier. Consolidates and issues their own BOL for shipment on a vessel. Does not own nor operate the vessel.
NVP	B	Net Present Value

Suggest New Words http://www.IDII.com/addword.htm

© 2004 Industrial Data & Information Inc.

NWPCA	Pa	National Wooden Pallet and Container Association - A national association with the goal of promoting the de- sign, manufacturer, distribution, recycling and sale of pallets, containers and reels. (P1)

Suggest New Words http://www.IDII.com/addword.htm

- O -

OAL	Conv	See Overall Length.
OAG	Std	Open Applications Group. Building standard OAGIS via XML for business applications. Focus is on standard data formats and interfacing for inter-operability between external vendor applications. See http://www.openapplications.org
OAW	Conv	See Overall Width.
OBC	Tr	On Board Computer. Many trucks have OBC to communicate with the company's dispatch and position tracking software.
OBI	B	Open Buying on the Internet. A project of the OBI Consortium (OBIC) in 1996 by major Fortune 500 companies in the US.
Obsolescence	B	The cost associated with inventory becoming obsolete. It can be extended to include such costs as disposal and storage. (KA)
Obsolete Stock	W	Stock held within an organization where there is no longer any reason for holding the stock.
OCR	C	Optical Character Recognition. Converts scanned document into text with expectations of highly accurate recognition rate. Depending upon the quality of the original documents, OCR may give you good results.
ODBC	C	Open DataBase Connectivity.
OE	B	Operational Excellence.
OE	C	Order Entry. Software where customer's quotes & sales orders are entered and managed. AKA OMS.
OEE	B Mfg	Operating Equipment Effectiveness. Business metric to measure the production line.

Suggest New Words http://www.IDII.com/addword.htm

© 2004 Industrial Data & Information Inc.

Glossary of Supply Chain Terminology

OEM	B	Original Equipment Manufacturer.
Off-Route Points	Tr	Points located off regular route highways of line-haul carriers, generally served only on irregular schedules. (Cnf)
OJT	B	On-the Job Training.
OLA	Org	Optical Laboratories Association. See http://www.ola-labs.org/
OLAP	C	Online Analytical Processing. A term for analyzing data into useful information. Other terms similar to this are BI, BPM, EPM, and EIS.
OLE	C	Object Linking and Embedding.
OLTP	C	On Line Transaction Processing.
OMB	B	Office of Management and Budget
OMG	Org Std	Object Management Group. See http://www.omg.org/
OMS	C	Order Management System. Software where customer's quotes & sales order are entered and managed. AKA OE.
On-Hand Balance	W	The recorded level of inventory in a distribution center. (KA)
ONS	C	Object Naming Service.
OO	C	Object Oriented.
OOS	B	Out Of Specification. When a sample of the product has been tested by QA/QC and found to be out-of-specification.
Open Systems	C	A phrase that represents a list of operating systems that are supported by many computer manufacturers. Windows, Linux, and Unix, are considered to be 'open systems'.
Opening Height	Pa	The vertical distance measured between decks, from the floor to the underside of the top deck, or from the floor to the top of the stringer notch. (P1)
Operating Authority	Tr	Routes, points, and types of traffic that may be served by carrier. Authority is granted by state or federal regulatory agencies. (Cnf)

Suggest New Words http://www.IDII.com/addword.htm

© 2004 Industrial Data & Information Inc.

Glossary of Supply Chain Terminology

Operating Efficiency	B	The percentage of actual output compared to standard output. (KA)
Operating Ratio	Tr	Comparison of carrier's operating expenses with gross receipts; income divided by expenses. Abbreviated as OR and O/R. (Cnf)
Operating System	C	Software that interfaces between the user and the hardware, and between applications and the hardware. The operating system allocates the resources of the computer, such as memory and processing time, to applications. Abbreviated OS. (KA)
Opportunity Costs	B	The expected returns of one activity, which are foregone in order to pursue other activities or projects. See Alternative Cost. (KA)
Optimization	B	Achieving the best possible solution or results with given resources and constraints. (KA)
OPWA	Org	Office Products Wholesalers Association. See http://www.opwa.org/
OR	B	Operations Research.
OR, O/R	Tr	See Operating Ratio.
Oral Toxicity	MSDS	Adverse effects resulting from taking a substance into the body via the mouth. Ordinarily used to denote effects in experimental animals. (HC)
Order Entry	C	Abbreviated OE. Software where customer's quotes & sales orders are entered and managed. AKA OMS.
Order Lead Time, Order Cycle Time	W	The total internal processing time necessary to transform a replenishment quantity into an order and for the transmission of that order to the recipient.

Suggest New Words http://www.IDII.com/addword.htm

© 2004 Industrial Data & Information Inc.

Glossary of Supply Chain Terminology

Order Picker	W	1) Lift truck which allows the warehouse worker to ride with the pallet and to pick from various levels. 2) A warehouse worker whose prime job is selection of orders. (KA)
Order Picking	W	Collecting items from a storage location or picking location to satisfy an order.
Ordering Costs	B	All of the costs associated with the task of preparing, transmitting, following up, and recording the receipt of an order. (KA)
O-Ring	Conv	Polyurethane bands polyurethane used to transmit drive power from roller to roller or spool to roller. (138-SP, 190-SP) (Fs)
OS	C	Operating System. E.G., Unix, AS/400, Windows 2000.
OSD	Tr W	See Over, Short, and Damaged.
OSF	Org	Open Software Foundation. See http://www.opengroup.org/
OSHA	MSDS Org	Occupational Safety and Health Administration. See http://www.osha.gov See also DOL and NIOSH, as OSHA collaborates with other government agencies for worker safety guidelines & regulations.
OSSSC	Org	Order Selection, Staging, & Storage Council. See MHIA.
OTS	B	Off The Shelf
Outbound Logistics	Tr	The portion of logistics that primarily involves the movement of materials and products from a company's production plant or storage warehouses. (KA)
Output Devices	C	Part of a computer system's harware used to communicate information. Examples are video monitors, printers, and modems. (KA)

Suggest New Words http://www.IDII.com/addword.htm

© 2004 Industrial Data & Information Inc.

Glossary of Supply Chain Terminology

Outside Dimension	Uom	Abbreviated OD. The exterior dimension of a container or package. In drums it is the diameter measured over the rolling hoops. (KA)
Outsourcing	B	Using a third-party manufacturer, carrier, or warehouse to perform functions formerly assigned to employees. (KA)
Overage	Tr W	Freight in excess over quantity believed to have been shipped, or more than quantity shown on shipping document. (Cnf)
Overall Height	Pa	The vertical distance measured from the floor to the top side of the top deck. (P1)
Overcharge	Tr	Where a shipper has been overcharged on its freight bill from the carrier. (WA)
Overhang	Pa	The distance the deck extends from the outer edge of the stringer or stringerboard; wing; lip. (P1)
Overhead Allocation	B	The portion of overhead cost that is applied by accountants to a particular product or service. (KA)
Overhead Cost	W	Those costs that are not directly related to warehousing and storage, which are still part of the total costs of a facility. These include janitorial services, heat, light, power, maintenance, depreciation, taxes, and insurance.(KA)
Overshipment	Pu W	A shipment containing more than originally ordered. (KA)
Over, Short, and Damaged	Tr W	A report issued at the warehouse when goods are damaged. Used to file a claim with the carrier. (Cnf)
Overall Length	Conv	The dimension outside of pulley to outside of pulley including belting or lagging, of any conveyor lengthwise. Abbreviated OAL. (Fs)
Overall Width	Conv	The dimension outside to outside of frame rails. Abbreviated OAW. (Fs)

Suggest New Words   http://www.IDII.com/addword.htm

© 2004 Industrial Data & Information Inc.

Overhead Drive	Conv	A drive assembly mounted over a conveyor, which allows clearance for the product. (Fs)
Owner-Operator	Tr	Drivers who own or operate their own trucks. May lease rig/driver to another carrier. (Cnf)
Owner's Risk	Tr	When owner of goods remains responsible during shipping, relieving carrier of part of risk. (Cnf)
Oxidizer	MSDS	A chemical other than a blasting agent or explosive that initiates or promotes combustion in other materials, thereby causing fire either of itself or through the release of oxygen or other gases. (HC)

- P -   - P -

P2P	C	Peer-to-Peer.
PAC	Mfg	Production Activity Control.
PAC	Org	Political Action Committees.
Pack Size	W	A collection of Units per Measure that describe how a Commodity is packaged. For example, Commodity ABC comes packaged in 24 Eaches per Case, and 50 Cases per Pallet.
Packaged Software	C	Complete software applications sold to a large market and not intended as custom software. (KA) See also COTS.
Package Stop	Conv	Any of various devices, either manual or mechanical, used to stop flow on a conveyor. (Fs)
Packing and Marking	W	The process of packaging goods for safe shipment and handling, and appropriately labeling the contents. (KA)
Packing List	W	List showing inventory that was packed in shipment. Prepared by warehouse personnel. Copy is sent to customer / consignee to help verify shipment received. Packing list detail can be also transmitted to customer / consignee via EDI Advance Ship Notice, if containers are uniquely coded.
Pallet	Pa	A portable, horizontal, rigid platform used as a base for assembling, storing, stacking, handling and transporting goods as a unit load, often equipped with a superstructure. (P1)
Pallet-Dimensions	Pa	When specifying pallet size, the stringer or stringerboard (block pallet) length is always expressed first; for example, a 48" x 40" pallet has a 48" stringer or stringerboard and 40" deckboards. (P1)

Suggest New Words   http://www.IDII.com/addword.htm

© 2004 Industrial Data & Information Inc.

Glossary of Supply Chain Terminology

Pallet Design System	Pa	A reliability based computer-aided design (CAD) program, for determining the safe load carrying capacity, performance, life and economy of wooden pallets. (P1)
Palletization	Pa W	System for shipping goods on pallets. Permits shipment of multiple units as one large unit. (Cnf)
Palletize	Pa W	To place material on a pallet in a prescribed area. (KA)
Palletizer	Pa W	A type of materials-handling device suing conveyor or robotics to position cubes or bags on a pallet. (KA)
Pallet Jack	Pa	Hand-propelled, wheeled platform, equipped with a lifting device for moving palletized unit loads. AKA low lift or pallet mover. (P1)
Pallet Life	Pa	The period during which the pallet remains useful, expressed in units of time or in the number of one-way trips. (P1)
Panel Deck Pallet	Pa	Pallet constructed with composite or structural panel top deck. (P1)
Parcel	Tr	In North America, parcel refers to shipments less than 150 lbs to be shipped via a small parcel carrier such as UPS, Fedex, Airborne, and USPO. Freight refers to shipments over 150 lbs to be shipped via LTL, TL, Ocean, Rail, or Air freight
Pareto's Law	B	A rule that says a relatively small number of products, sales, or activities compromise a large percentage of the total. First described by Italian sociologist Vilfredo Pareto. Also called the 80/20 rule. See ABC Management. (KA)
PARM	B	Parameter. Common abbreviation used in business and in computer programming.
Partial Four-Way Stringer Pallet	Pa	A pallet with notched stringers. (P1)

Suggest New Words   http://www.IDII.com/addword.htm

© 2004 Industrial Data & Information Inc.

Glossary of Supply Chain Terminology

Parts Conveyor	Conv	A conveyor used to catch and transport small parts, stampings, or scrap away from production machinery to hoppers, drums, or other operations. (PC, PCA, PCX, PCH) (Fs)
PASBA	Org	Procurement And Supply-Chain Benchmarking Association. See http://www.pasba.com/
Payload	Tr	1) Carried cargo. 2) The net weight of the cargo. (Cnf)
PBR	W	Push back rack.
PC	C	Personal Computer.
PC	Mfg	Production Control
PCB	C	Printed Circuit Board
PCS	Mfg	Process Control System (EM)
PCT	B	Percent. Common abbreviation used in business and in computer programming.
P&D	Tr	Pickup and Delivery.
PDA	C	Personal Digital Assistant or Personal Data Assistant.
PDF	C	Portable Document File. A common computerized document format for keeping text, fonts, and images correctly paginated for presentation. Developed by Adobe.
PDF417	Std	2D stacked barcode symbology where one may encode a large amount of information within the PDF417 barcode.
PDM	W	Product Data Management. Software module that is focused on proper setup, maintenance, and changes to the life of the product.
PDS	Pa	See Pallet Design System.
PDT	C W	Portable data terminals. A rugged hand-held computer with the computing power of a stationary computer. Normally equipped with RF (radio frequency) and an antenna.
PEL's	MSDS	See Permissible Exposure Limits.

Suggest New Words   http://www.IDII.com/addword.htm

© 2004 Industrial Data & Information Inc.

Glossary of Supply Chain Terminology

Term	Cat	Definition
PERA	Org	Production Engine Remanufacturers Association. See http://www.pera.org/
Per Diem	Tr	Latin term meaning "by the day." Daily charge for use of railcars. (Cnf)
Perishable Freight	Tr	Commodities subject to rapid deterioration or decay, which require special protective services such as refrigeration or heating. (Cnf)
Permissible Exposure Limits	MSDS	Permissible Exposure Limits (PEL's) are OSHA's legal exposure limits. (HC)
Permit	Tr	Authority granted to contract carriers and forwarders by the ICC to operate in interstate commerce. (Cnf)
Perpetual Inventory System	W	An inventory control system where a running record is kept of the amount of stock held for each item. Whenever an order is filled, the withdrawal is logged and the result compared with the re-order point for any necessary re-order action.
PERT	B	Program and Evaluation Review Technique.
PF&D	W	Personal fatigue, and delay times. This is expressed as a percentage of time allowed for completing a task. (KA)
PFR	W	Pallet Flow Rack.
PH	MSDS	A symbol relating the hydrogen ion (H+) concentration of that of a given standard solution. A pH of 7 is neutral. Numbers from 7 to 14 indicate greater alkalinity. Numbers from 7 to 0 indicate greater acidity. (HC)
Phantom Bill	M	A bill of material (BOM) for intermediate manufacturing steps. The results of the phantom bill are used in the next step of manufacturing.
Photo Cell	Conv	Electrical device used to sense product location. (Fs)

Suggest New Words http://www.IDII.com/addword.htm

© 2004 Industrial Data & Information Inc.

Glossary of Supply Chain Terminology

Term	Src	Definition
Physical Distribution	B	A term that applies to the logistics activities that occur between the end of the production line and the final user. It includes traffic, packaging, materials handling, warehousing, order entry, customer service, inventory control, and forecasting. (KA)
Physical Hazard	MSDS	A chemical for which there is scientifically valid evidence that it is a combustible liquid, a compressed gas, explosive, flammable, an organic peroxide, an oxidizer, pyrophoric, unstable (reactive) or water-reactive. (HC)
Physical Inventory	W	A physical count of every item located within the warehouse. AKA Annual Inventory. Cycle counts can replace physical counts. (KA)
PI	W	Physical Inventory
Pick-and-Pass	W	Picking technique used with pick zones, flow racks & conveyors. One picker will pull products into tote (or other container). Tote will be passed to next zone for picker to pull and put products into same tote. This pick-and-pass continues until all zones are completed and then tote is taken away on a take-away conveyor.
Pick Time	W	The amount of time it takes a worker to select and document an item. (KA)
Pick-to-Carton	W	Picking technique where picker is directed to put pulled product directly into a licensed plated container (a "carton" normally). Picker may be picking for multiple orders during one trip, by using multiple cartons. When finished the picker puts the cartons on a take away conveyor for the shipping station.
Pick Face, Picking Face	W	See Pick Slot.

Suggest New Words http://www.IDII.com/addword.htm

© 2004 Industrial Data & Information Inc.

Glossary of Supply Chain Terminology

Pick-to-Light	W	Picking technique used where pickers are directed by lights and/or digital displays at each bin. Very effective technique for high volume piece picking. Requires the WMS to be interfaced to Pick-To-Light MHE interface.
Pick List	W	An output from a warehouse management system designating those items, by item number, description and quantity, to be picked from stock to satisfy customer demand.
Pick Slot	W	The location where inventory is stored that is dedicated or picking. The pick slot is commonly replenished via a minimum trigger point to initiate replenishment. The terms forward location, primary location, and picking face are synonyms for this term.
Picker	W	Term used for the person assigned the job of locating and removing stock from storage locations. See Order Picker. (KA)
Picket Fence	Std	A barcode printed horizontally so that the bars look like a picket fence.
PIDA	Org	Pet Industry Distributors Association. See http://www.pida.org/
Piggyback	Tr	A form of intermodal transportation where trailers/containers are carried on railcars. (Cnf)
Pilferage	B	1) The felonious act of breaking into containers and removing property. 2) Petty theft of merchandise. (KA)
Ping	C	Packet Inter-Network Groper. An easy way to test if a computer, hub, router, switch, or network printer is ready to be utilized by users on a network. Ping sends a packet to a designated address and waits for a response.

Suggest New Words http://www.IDII.com/addword.htm

© 2004 Industrial Data & Information Inc.

Pinwheel Pattern	W	A method of storing merchandise on a pallet in a pattern to arrange items of unequal width or length. See Brick Pattern and Row Pattern. (KA)
PIP	C	Partner Interface Process. RosettaNet has developed an XML standard, using PIPs & DTDs. PIPs describe the message mapping, message flow, and process interaction between trading partners. All PIP message schemas are defined using Document Type Definitions (DTDs).
Pipeline Stock	B	Inventory within the pipeline, including in-transit inventory as well as inventory positioned in distribution centers. (KA)
Pivot Plate	Conv	The gusset that attaches the conveyor to the support leg. (Fs)
PKI	C	Public Key Infrastructure. PKI is a specification using private cryptographic key pairs from a trusted authority. The end result is that PKI users can privately & securely exchange data and money on public networks, such as the internet.
P&L	B	Profit and Loss.
Planning Horizon	Mfg	The amount of time a schedule or forecast extends into the future. (KA)
Plastisol Coating	Conv	Poly-vinyl chloride (PVC) covering for roller tubes to prevent product damage or marking. Usually (#70 durometer) green or (#90 durometer) red in color. (Fs)
PLC	B	Public Limited Company.
PLC	C	Programmable Logic Controller. Automates and controls either material handling or Manufacturing automation equipment. E.G., diverters controlled by PLC.

Suggest New Words http://www.IDII.com/addword.htm

© 2004 Industrial Data & Information Inc.

Glossary of Supply Chain Terminology

PLM	B	Product Lifecycle Management. A way to manage product data from the concept stage until a product sunsets years later. Software vendors are promoting this valuable concept.
Plow	Conv	A device positioned across the path of a conveyor at the correct angle to discharge or deflect objects. (Fs)
PLU	B	Price Look Up.
PM	B	Preventive Maintenance. Equipment needs scheduled PM for cleaning and replacement of maintenance components.
PMAC	Org	Purchasing Management Association of Canada. See http://www.pmac.ca/
PMI	Org	Project Management Institute. With over 100,000 members worldwide, PMI is the leading professional association in the area of Project Management. See http://www.pmi.org
PML	C	Physical Markup Language. A data transfer specification based in XML. The Auto ID Center is promoting the PML with ePC and ONS.
PN, P/N	B	Part Number. A code to uniquely identify a part/product. See UPC.
PO	B Pu	Purchase Order. Document sent to a vendor indicating goods or services to be purchased at a specific price and terms for indicated quantities. See EDI 850.
POD	Tr	See Proof of Delivery.
POE	Tr	See Port of Entry.
POI	B	Point Of Installation (EM)
Point of Origin	Tr	The station at which a shipment is received from the shipper by a transportation line. (Cnf)

Suggest New Words http://www.IDII.com/addword.htm

© 2004 Industrial Data & Information Inc.

Glossary of Supply Chain Terminology

Term	Cat	Definition
Polymerization	MSDS	A chemical reaction in which one or more small molecules combine to form larger molecules at a rate that releases large amounts of energy. If hazardous polymerization can occur with a given material, the MSDS will usually list conditions which could start the reaction. In most cases the material contains a polymerization inhibitor, which if used up, is no longer capable of preventing a reaction. (HC)
Poly-Tier Support	Conv	Supporting members capable of supporting more than one level of conveyor at a time. Each tier has vertical adjustment for leveling the conveyor. (Fs)
Pooling Agreement	Tr	The dividing of revenue/business among two or more carriers in accordance with previous contracts/agreements. (Cnf)
Pop-Out Roller	Conv	A roller, normally placed on the ends of a belt conveyor used to aid in transfer, and set in a wide groove to allow it to eject if an object comes between it and the belt. (Fs)
Portable Conveyor	Conv	Any type of transportable conveyor, usually having supports that provide mobility. (Fs)
Portable Plate	W	A loading ramp that can be removed to any loading position on the dock. (KA)
Portable Support	Conv	Supporting members that provide conveyor mobility by use of casters or wheels. (Fs)
Port of Entry	Tr	A port at which foreign goods are admitted into the receiving country. (WA) Abbreviated POE.
POS	B	Point of Sale.

Suggest New Words http://www.IDII.com/addword.htm

© 2004 Industrial Data & Information Inc.

Glossary of Supply Chain Terminology

Term	Cat	Definition
Post Audit	W	A study conducted of a new warehouse, fleet, or equipment to ascertain how well it is performing in relation to the proposal and financial analyses used to originally justify it. (KA)
Postponement	B	The delay of the final packaging, branding, formulation, or commitment of a product until the last possible moment. The practice reduces speculation or risk by delaying differentiation of the product until it is purchased. (KA)
Postponement of Final Assembly	W	The delay of final assembly until a firm customer order is received. Often, common parts and components can be produced, shipped, and inventoried at lower cost and risk than that associated with completed products. (KA)
Positive Crowned Pulley	Conv	A pulley that tapers equally from both ends toward the center, the diameter being the greatest at the center. The crown aids in belt tracking. (Fs)
Postnet	Std	Barcode symbology used on mail for the US Post Office to encode zip codes (postal codes).
Post Pallet	Pa	A pallet fitted with posts or blocks between the decks or beneath the top deck. See Block Pallet. (P1)
POU	B	Point Of Use (EM)
Power Belt Curve	Conv	A curve conveyor that utilizes a belt, driven by tapered pulleys. (Fs)
Power Conveyor	Conv	Any type of conveyor that requires power to move its load. (Fs)
Power Feeder	Conv	A driven length of belt conveyor, normally used to move product horizontally onto an incline conveyor. (Fs)
PP	Gov	Past Performance.
PPB	MSDS	Parts Per Billion. Parts of vapor or gas per billion parts of contaminated air by volume. (HC)
PPM	C	Pages Per Minute

Suggest New Words http://www.IDII.com/addword.htm

© 2004 Industrial Data & Information Inc.

Glossary of Supply Chain Terminology

Term	Cat	Definition
PPM	MSDS	Parts Per Million. Parts of vapor or gas per million parts of contaminated air by volume. (HC)
Preemption	Gov	The principle – derived from the Supremacy Clause of the Constitution – that a federal law supersedes any inconsistent state law or regulation. (WA)
Pressure Label	W	A price ticket or other information ticket that can be affixed to merchandise by pressing it on. (KA)
Pressure Roller	Conv	A roller used for holding the driving belt in contact with the load carrying rollers in a belt driven live roller conveyor.
Preventive Maintenance	B	Actions taken to reduce break-downs in machinery and equipment, such as basic cleanliness and replacement or adjustment of parts. (KA)
Primary Location	W	See Pick Slot.
Privacy of Contract	Tr	Contract term meaning the parties legally bound and able to recover under the contract. (WA)
Private Carrier	Tr	A company that provides its own transportation, either through leased or owned equipment. Private carriers are allowed to transport goods for subsidiaries and to backhaul products from non-affiliated companies. (KA)
Private Warehouse	W	A warehouse operated by the owner of the goods stored there. A private warehouse can be an owned or leased facility. (KA)
PRM	B	Partnership Relationship Management. See also CRM, ERM, and SRM.
PRM	B	Price-Revenue Management. SCM must take into effect both CRM & PRM issues.

Suggest New Words http://www.IDII.com/addword.htm

© 2004 Industrial Data & Information Inc.

Glossary of Supply Chain Terminology

PRO	B	Price Revenue Optimization. A methodology or software that determines which products and services should be price at, in order to maximize profits.
Procurement	Pu	The first phase of the supply process. Procurement includes all activities from the selection of vendors to the purchase and transport of raw materials to the point of manufacture, reprocessing, or repackaging. (KA)
Product Footprint	Conv	The surface of the product that comes in contact with the belt rollers, or wheels of the conveyor. (Fs)
Production Scheduling	Mfg	The process of scheduling all activities and materials required for manufacturing. The organization of material flows into the production setting in a timely manner. (KA)
Product Life Cycle	B	The stages a product evolves through during the period that it is marketed. The four general phases are introduction, growth, maturity, and marketing. (KA)
Product Line	B	A group of similar products that are aggregated for planning and marketing. (KA)
PROM	C	Programmable Read Only Memory
Promotional Product	B	A product that is the subject of a campaign to stimulate sales by price reductions or other incentives. This process may cause a wide fluctuation in demand. AKA Promo. (KA)
Pro Number	Tr W	A unique number assigned to a freight shipment by the carrier. This Pro Number is used on freight bill and bill of ladings.

Suggest New Words   http://www.IDII.com/addword.htm

© 2004 Industrial Data & Information Inc.

Glossary of Supply Chain Terminology

Term	Type	Definition
Proof Of Delivery	Tr	Copy of waybill signed by consignee at time of delivery as receipt. Abbreviated POD. (Cnf)
Proportional Weight	Tr	Lower than normal rate on segment of through movement to encourage traffic or capture competitive traffic. May be percentage of standard rate or flat rate that is lower between given points. (Cnf)
Pro Se	Tr	On one's own behalf. (WA)
Protocol	C	A method to communicate data between software applications on one or more computer systems.
Product Footprint	Conv	The surface of the product that comes in contact with the belt rollers, or wheels of the conveyor. (Fs)
Proximity Switch	Conv	An electrical device to control some function of a powered conveyor. (Fs)
PSA	B	Product Service Agreements (EM)
PSCM	Pu	Procurement In Supply Chain Management (EM)
PSDA	Org	Pacific Southwest Distributors Association. See http://www.asaonline.org/
PSI	Uom	Pounds per Square Inch.
PTD	B	Period-To-Date. Common abbreviation used in business and in computer programming.
PTD	B	Profitable To Deliver. Order commitment based upon CTD and the profitability that includes the execution side (transportation, distribution, logistics). See also PTP.
PTL	W	Pick To Light technology uses displays at each location that light up to show where to pick. Picker goes to this location, selects quantity displayed and pushes a button to confirm the pick. Used in high volume items. Cost is estimated to be $200 per light position in 2001.

Suggest New Words http://www.IDII.com/addword.htm

© 2004 Industrial Data & Information Inc.

Glossary of Supply Chain Terminology

Term	Cat	Definition
PTP	B	Profitable To Promise. In addition to ATP and CTP, the PTP analyzes profitability of all factors involved in this order fulfillment.
Public Warehouse	W	See also Contract Warehouse and Third Party Warehouse.
PU&D, PUD	Tr	Pick-Up and Delivery.
Pull Distribution	B	System in which retail demand stimulates inventory and transportation flows. (KA)
Pulley	Conv	A wheel, usually cylindrical, but polygonal in cross section with its center bored for mounting on a shaft. (Fs)
Purchase Requisition	Pu	An intracompany form that instructs the firm's purchasing agent to acquire merchandise or supplies. (KA)
Purchasing Agent	Pu	The individual authorized to purchase goods and services for a company. (KA)
Purchasing Costs	Pu	All costs associated with the purchasing activities. These costs include personnel, office supplies, and communication charges. (KA)
Push-Back Rack	W	Rack system that allows palletized product to be stored by being pushed up an inclined ramp. This allows for deep pallet storage.
Push Button System	Conv	An electrical device that operates a magnetic starter. (Fs)
Push Distribution	B	A system in which distribution centers and retail points are stocked in anticipation of demand. (KA)
Pusher	Conv	A device, normally air powered, for diverting product from one conveyor line to another line, chute, etc. (Fs)
Purchase Order	B	Document sent to a vendor indicating goods or services to be purchased at a specific price and terms for indicated quantities. AKA PO. See EDI 850.
Putaway	W	The movement of material from the point of receipt to a storage area. (KA)

Suggest New Words http://www.IDII.com/addword.htm

© 2004 Industrial Data & Information Inc.

Glossary of Supply Chain Terminology

- Q -		- Q -
QA	W	See Quality Assurance/Quality Control.
QC	W	See Quality Assurance/Quality Control.
QTD	B	Quarter-To-Date. Common abbreviation used in business and in computer programming.
QTY	B	Quantity. Common abbreviation used in business and in computer programming.
Quality Circle	B	A team of workers that meets to solve quality control problems. (KA)
Quantity Discount	B	A reduction in the purchase price that results from increasing of the quantity or value of the overall order. (KA)
Quarantine	W	The isolation of goods or materials until they can be checked for quality or conformance with all required standards. (KA)
Queue	B	The line formed by loads, items or people while waiting. (KA)
Quid Pro Quo	Tr	The act of exchanging one thing for another. (WA)
Quotas	Gov Tr	Many governments have established quotas of limiting imports by class of goods or country of origin. Sometimes importing countries require issuance of licenses before U.S. companies may ship to them. (Cnf)
QoS	B C	Quality of Service. Describes fast network speeds with quality of service option
QR	B	Quick Response.
QR	W	Quality Review. Check procedure on randomly selected components and/or finished goods product. See also QA, QC.
QRP	B	Quick Response Program.

Suggest New Words http://www.IDII.com/addword.htm

© 2004 Industrial Data & Information Inc.

Glossary of Supply Chain Terminology

QS	W	Quality Standard. See Automotive Industry Action Group requirements. (EM)
Quality	Pa	Consistent performance of a uniform product meeting the customer's needs for economy and function. (P1)
Quality Assurance/Quality Control	W	Processes of inspecting merchandise being received or shipped to ensure that the merchandise is of adequate quality and that the case content specification is accurate.
Quarantine	W	An inventory status indicating that this product is not available for inventory allocation but is in process of inspection for QA or another reason. See also QA and QC.
Query	C	In data processing, the process of asking a database program for a list of records that fit certain criteria. (KA)
Quiet Zone	Std	A blank area around the printed barcode this is required. This quiet zone size is small and varies upon the barcode symbology.

Suggest New Words http://www.IDII.com/addword.htm

© 2004 Industrial Data & Information Inc.

Glossary of Supply Chain Terminology

- R -		- R -
R2R	B	Reseller-to-Reseller. Similar to B2B, but each party involved is a reseller. Therefore, the product or service being sold by the reseller (R2R) is purchased by another reseller for resale.
Rack	Conv	As in a curve, to bend the frame racks down in order that the discharge elevation is lower than the infeed elevation. (Fs)
Rack	W	A structured storage system (single-level or multi-level) that is used to support high stacking of single items or palletized loads. (KA)
Rack-Supported Building	W	A warehouse in which the storage rack functions as the structural support for the roof. (KA)
Racked Across Deckboards	Pa	Output from PDS program describing the maximum load carrying capacity and deflection of a pallet where the rack frame supports the pallet only at the ends of the deckboards. (P1)
Racked Across Stringers	Pa	Output from PDS program describing the maximum load carrying capacity and deflection of a pallet where the rack frame supports the pallet only at the ends of the stringers or stringer boards. (P1)
Radio Frequency	W	A system of mobile devices used to issue tasks, edit entered information and confirm the completion of tasks using either laser scanners or terminal entry. RF allows operations personnel to function in a "point-of-work" mode.
Radio Interrogation Signal	C	The signal produced by the reader for RFID system that causes the tag to transmit the ID.

Suggest New Words   http://www.IDII.com/addword.htm

© 2004 Industrial Data & Information Inc.

Glossary of Supply Chain Terminology

RAM	C	Random Access Memory. All computers require memory, which is called RAM. Memory in the computer or needed in the computer is expressed in MB (Megabytes).
Ramp	W	An inclined roadway that connects levels in the warehouse. (KA)
Random Storage	W	One or more areas of the warehouse that are designed for random storage. Bins in a random storage area are not pre-assigned, but the WMS is allowed to fully evaluate what inventory to locate in these random storage bins.
Rate	Tr	Established shipping charge for movement of goods. In interstate transportation, price/rate is approved by ICC. Intrastate prices are approved by public service commission or similar body. (Cnf) See also FAK.
Rate Basis	Tr	Formula of specific factors/elements that control making of rate. (Cnf)
Rate Bureaus	Tr	A carrier group that assembles to establish joint rates, to divide joint revenues and claim liabilities, and to publish tariffs. (WA)
Rate Files	Tr	The collection of published transportation prices or tariffs. (KA)
Rate War	Tr	When carriers cut rates in an effort to secure tonnage. Can occur in all commodities. (Cnf)
Raw Material	Mfg	Goods awaiting conversion into manufactured products or components. (KA)
RCCP	Mfg	Rough-Cut Capacity Planning.
RCPT	B	Receipt. Common abbreviation used in business and in computer programming.
RCRA	MSDS	Resource Conservation and Recovery Act, administered by the EPA.
R&D	B	Research & Development.

Suggest New Words http://www.IDII.com/addword.htm

© 2004 Industrial Data & Information Inc.

Glossary of Supply Chain Terminology

RDC	W	Retail Distribution Center. A DC replenishing products directly to the stores. See Distribution Center.
Reach Truck	W	A forklift with the extended ability to reach forward significantly. Very useful for getting or retrieving the deeper pallet in a double deep pallet bin.
Real Time	W	Real time means that the new or updated information is instantaneously saved in the database.
Reactivity	MSDS	A description of the tendency of a substance to undergo chemical reaction with the release of energy. Undesirable effects such as pressure build-up, temperature increase, and formation of noxious, toxic or corrosive byproducts may occur because of the reactivity of a substance by heating, burning, direct contact with other materials, or other conditions in use or in storage. (HC)
Reasonable Care	W	The extent to which a warehouse operator is liable for goods. As defined in section 7-240-1 of the Uniform Commercial Code: A warehouseman is liable for damages or injury to the goods caused by his failure to exercise such car in regard to them as a reasonably careful person would exercise under like circumstances, but unless otherwise agreed he is not liable for damages that could not have been avoided by the exercise of such care. (KA)
Rebate	Tr	Unlawful practice in which carrier returns part of transportation charge to shipper. Done to encourage shipper to use same carrier again. (Cnf)
REC	B	Record. Common abbreviation used in business and in computer programming. Usually refers to a specific record in a database table.

Suggest New Words http://www.IDII.com/addword.htm

© 2004 Industrial Data & Information Inc.

Glossary of Supply Chain Terminology

Receiving End	Conv	The end of a conveyor from which the products move toward you. (Fs)
Receiving Report	W	A record of the condition in which merchandise arrived. (KA)
Receiving Tally	W	The warehouse receiver's independent listing of goods unloaded from an inbound vehicle, sometimes prepared on a blind basis to ensure accuracy. (KA)
Reciprocity	Tr	1. An exchange of rights. In motor transportation, may involve granting equal rights to vehicles of several states in which reciprocity agreements are in effect. 2. To give preference in buying to vendors who are customers of buying company. (Cnf)
Reconsign-ment	Tr	1) A change (usually requested by the shipper) in the routing, or destination, of a shipment that is already in transit but does not require a new freight bill. 2) A fee for the latter. (See also DIVERSION) (Cnf)
Recycling	Pa	A pallet, container or reel that has been used, discarded, salvaged, repaired and which passes through a cycle again. (P1)
REDINET	C	EDI network originally developed by Control Data Corporation; operated by Sterling Commerce's Network Services Group since June 1991.
Redundancy	C	Backup capabilities and duplications of a system designed to reduce or eliminate downtime due to breakdowns. (KA)
Reefer	Tr	A refrigerated flat bed trailer that hauls perishables.
REF	B	Reference. Common abbreviation used in business and in computer programming.

Suggest New Words http://www.IDII.com/addword.htm

© 2004 Industrial Data & Information Inc.

Glossary of Supply Chain Terminology

Term	Cat	Definition
Refrigerated Container	Tr	An insulated container that provides temperature-controlled environment to protect perishable materials. (KA)
Refrigerated Warehouse	W	A warehouse that provides refrigeration and temperature control for perishable products. (KA)
Release	W	The authorization to ship material. (KA)
Released Rates	Tr	Rates based upon the shipment's value. The maximum carrier liability for damage is less than the full value, and in return the carrier offers a lower rate. (WA)
Renewal Storage	W	The rebilling fee (usually monthly) for products stored in a public warehouse. (KA)
Rental Pallet	Pa	A pallet owned by a third party, different from the actual pallet user. (P1)
Re-Order Point	Pu	When on-hand quantity of the specified product decreases to this re-order point or less, it is time to reorder.
Repack	W	Task that takes product in one configuration and is repackaged into another. 8 packs into 12 packs.
Repair	Pa	To remake in order to use again. (P1)
Replacement Cost	B	The current fair-market price to purchase an equivalent asset. (KA)
Replenishment	W	The task initiated to fill a picking location. Typically a replenishment task can be specific to an item's replenishment quantity and unit of measure.
REQ	Pu	Purchase Requisition.
Reserved Inventory	B	Inventory that is held in reserve for one or more customers.(KA)
Restricted Articles	Tr	Commodities that can be handled only under certain specific conditions. (Cnf)
Return Idler	Conv	A roller that supports the return run of the belt. (Fs)
Return Receipt	Tr	A form sent to the shipper, after a consignee has received a shipment, that indicates delivery has been made. (KA)

Suggest New Words http://www.IDII.com/addword.htm

© 2004 Industrial Data & Information Inc.

Glossary of Supply Chain Terminology

Returnable Pallet, Reusable Pallet	Pa	A pallet designed to be used for more than one trip. (P1)
Reversible	Conv	A conveyor that is designed to move product in either direction. (Fs)
Reverse Logistics	B W	The requirement to plan the flow of surplus or unwanted material or equipment back through the supply chain after meeting customer demand.
Rewarehousing	W	The process of calculating the best slotting positions for inventory and moving inventory to those optimized bins. This is done to optimize space utilization and decrease deadheading travel time.
RF	W	See Radio Frequency.
RF/DC	C	Radio Frequency Data Communications.
RFDC	W	Radio Frequency Data Collection (EM)
RFI	B	Request For Information. A request for a vendor to provide information. Not as intensive as a RFP, but may provide detailed functional answers to what a solution can provide.
RFID	W	Radio Frequency Identification. Usually refers to RFID tags and readers.
RFP	B	Request For Proposal. A request for a vendor to do a full proposal including costs, references, project plans (implementation), and detailed functional answers from a vendor to a prospective client.
RFQ	B	Request For Quotation. A cost proposal for a given service or product. High-level summaries are given on functions.
RICI	Org	Remanufacturing Industries Council International. See http://www.reman.org/
RICO	Gov	Racketeer Influenced and Corrupt Organizations Act.
RLEC	Org	Reverse Logistics Executive Council. See http://www.rlec.org/

Suggest New Words   http://www.IDII.com/addword.htm

© 2004 Industrial Data & Information Inc.

Glossary of Supply Chain Terminology

RM	B Tr	Revenue Management. Determination of total cost, labor, and how to determine pricing for revenue analysis. Being done especially with Cargo Revenue Management.
RMA	B	Return Merchandise Authorization.
RNIF	C	RosettaNet Implementation Framework. RNIF is a set of specifications utilized to implement message exchange between trading partners. RosettaNet is an XML standard, using PIPs & DTDs.
RO	C	Read Only
ROA	B	Return on Assets.
Roaming	C	Allows users free travel from one access point (AP) area of coverage to another with no loss in connectivity. Utilized in RF and Bluetooth technologies.
ROE	B	Return on Equity.
ROI	B	Return On Investment. Calculations that result in a measurement of what a company will gain from an investment.
Roller	Conv	A round part free to revolve about its outer surface. The face may be straight, tapered or crowned. Rollers may also serve as the rolling support for the load being conveyed. (Fs)
Roller Bed	Conv	A series of rollers used to support a conveying medium. (Fs)
Roller Centers	Conv	The distance measured along the carrying run of a conveyor from the center of one roller to the center of the next roller. (Fs)
Roller Conveyor	Conv	A series of rollers supported in a frame over which objects are advanced manually, by gravity or by power. (Fs)
Roll-On / Roll-Off	Tr	A feature in specially constructed vessels permitting road vehicles to drive on/off a vessel in loading/discharging ports. (Cnf)

Suggest New Words   http://www.IDII.com/addword.htm

© 2004 Industrial Data & Information Inc.

Glossary of Supply Chain Terminology

ROM	C	Read Only Memory
ROO	B	Return On Opportunity. Calculations that result in a measurement of what a company will gain from an opportunity.
ROP	Pu	Re-order Point. When on-hand quantity of the specified product decreases to this re-order point or less, it is time to reorder.
RO/RO	Tr	See Roll-On / Roll-Off.
RosettaNet	Org	RosettaNet. Building standard interface specifications for commerce based on XML. Similar to OAG & OAGIS for inter-operability between vendor applications. See http://www.rosettanet.org
Route	Tr	1. Course/direction that shipment moves. 2. To designate course/direction shipment shall move. 3. Carrier(s) with junction points over which shipment moves. (Cnf)
Routing	Tr	1. Process of determining how shipment will move between origin and destination. Routing information includes designation of carrier(s) involved, actual route of carrier, and estimated time en route. 2. Right of shipper to determine carriers, routes and points for transfer on TL and CL shipments. (Cnf)
RPC	C	Remote Procedure Call.
RPC	W	Reusable Plastic Container.
RPI	Mfg	Raw Products Inventory
RPM	B	Real-Time Performance Management. Business metrics to measure performance in real-time.
RPM	C	Revolutions Per Minute
RPT	B	Report. Common abbreviation used in business and in computer programming.
RSN	B	Reason. Common abbreviation used in business and in computer programming.

Suggest New Words http://www.IDII.com/addword.htm

© 2004 Industrial Data & Information Inc.

RSS	Std	Reduced Space Symbology code from the UCC is 14 digits. A technique used to conserve space in the barcode for UCC/EAN. See also GTIN and ePC.
RT	C	Real Time
RTLS	Tr W	Real Time Locating System. Usually in reference to a WMS or YMS ability to move-track-locate inventory.

Suggest New Words http://www.IDII.com/addword.htm

© 2004 Industrial Data & Information Inc.

- S -		- S -
Safety Stock	Pu	A minimum quantity of stock carried in inventory in addition to the forecasted customer requirements. Safety stock is meant to provide sufficient stock for emergencies, unanticipated demand, or unforeseen delays. (KA)
SAL	Org	Smart Active Label (SAL) Consortium. See http://www.sal-c.org/
Salable Goods	B	Products authorized for sale to customers. (KA)
SAMI	Pu	Supplier Assistance in Managing Inventories (EM)
SAP	Org	SAP Inc. A worldwide provider of ERP software. See http://www.sap.com/
SARA Title III	MSDS	Title III of the Superfund Amendments and Reauthorization Act of 1986, also known as the Emergency Planning and Community Right-To-Know Act. It requires extensive submission of information about hazardous chemicals to EPA, states, and local communities, and establishes a national program of emergency planning. Administered by EPA. (HC)
SAT	B	Site Acceptance Test. See also Factory Acceptance Test (FAT).
Satellite	Tr	1. A local servicing terminal. Also called "city," "group," or "end-of-the-line" terminals. 2. An accounting designation for a specific service area; not a regular freight terminal location. (Cnf)
SBA	Org	Small Business Administration. See http://www.sba.gov/
SC	B	Supply Chain.
SC	Conv	See Sortation Conveyor.

Suggest New Words http://www.IDII.com/addword.htm

© 2004 Industrial Data & Information Inc.

Glossary of Supply Chain Terminology

Term	Type	Definition
SCA	Org	Shipbuilders Council of America. See http://www.shipbuilders.org/
SCAC	Std Tr	Standard Carrier Alpha Code is a unique code to identify transportation companies. The Standard Carrier Alpha Code is the recognized transportation company identification code used in the American National Standards Institute (ANSI) Accredited Standards Committee (ASC) X12 and United Nations EDIFACT approved electronic data interchange (EDI) transaction sets. The NMFTA assigns SCAC codes for all motor carriers. See http://www.nmfta.org
Scanning Equipment	C	Machines used to read and process transactions by reading encoded information and accounting for the transactions. (KA)
SCC	Org	Supply Chain Council. See http://www.supply-chain.org/
SCC, SCC-14	Std W	Shipping Container Code consisting of 14 digits. This UCC SCC-14 code is frequently put on a pallet or container label. The first two o digits are the UCC application identifier. The next digit is the packaging level indicator. The next seven digits is the manufacturer identification number. The next five digits represent the manufacturer assigned product number. The last digit is a modulus 10 check digit. See also UCC 128.
SCE	W	Supply Chain Executions. Refers to the "execution" side of fulfilling the customer order and supporting functions to do so. Commonly includes warehouse, transportation, data interchange.

Suggest New Words http://www.IDII.com/addword.htm

© 2004 Industrial Data & Information Inc.

Glossary of Supply Chain Terminology

SCES	C W	Supply Chain Execution Systems. Software and processes to enable then execution side of fulfilling customer orders and supporting functions. Includes warehouse operations, transportation, with heavy emphasis on inventory & order entry.
Scheduled Downtime	Mfg	A planned shutdown of operations to perform maintenance or react to decrease demand. (KA)
SCL	Org	Canadian Association of Supply Chain & Logistics Management. See http://www.infochain.org/
SCM	B	See Supply Chain Management.
SCO	B	Supply Chain Optimization.
SCOR	B	Supply Chain Operations Reference-Model.
SCP	B	Supply Chain Planning.
SCPM	C	Supply Chain Process Management. SCPM provides alerting, alert response logic, inventory visibility, and KPI. SCPM includes SCEM plus visibility, KPI, and non-compliance event management.
SCRA	Org	Specialized Carriers and Rigging Association. See http://www.scranet.org/
Scrap	Mfg	Excess material, or material that does not meet the required standards, and is unable to be re-worked economically. (KA)
Scrap Factor	Mfg	The percentage of raw materials or components included in gross requirements and expected to be lost in the manufacturing process. (KA)
SCSI	C	Small Computer System Interface. Hard disks and other devices have SCSI interfaces. Other types of interfaces are common as well. These interfaces let the computer communicate (read / write) to the devices (I.E., Hard disk).

Suggest New Words   http://www.IDII.com/addword.htm

© 2004 Industrial Data & Information Inc.

Glossary of Supply Chain Terminology

SCU	W	Speech Control Unit. The Speech Control Unit is a computer that connects to the WMS for exchange of information. The SCU then gives individual instructions to the SDT via a wireless network. See also SDT.
SCV	B	Supply Chain Visibility.
SDR	B	See Special Drawing Rights.
SDT	W	Speech Data Terminal. In voice-activated devices, mobile workers use small, unobtrusive wearable computers called Speech Data Terminals with an attached headset. This would be used in a pick-to-voice environment. See SCU.
Seal	Tr	A lockable numbered metal strip applied to the door of a retail car, truck, or container. A broken seal indicates that the door has been opened.(KA)
Seasonal Index	Pu	A number used to adjust forecasts and optimal inventory calculations for seasonal fluctuations. (KA)
Seasonal Inventory	Pu W	Inventory held to meet seasonal demand. (KA)
SED	Tr	See Shipper's Export Declaration.
SEDA	Org	Safety Equipment Distributors Association. See http://www.safetycentral.org/
SEF	C	Standard Exchange Format. Other ways to exchange data include ODBC, EDI, and XML.
Semi	Tr	Slang term for semi-trailer. Also used loosely in referring to tractor-trailer combination. See Semi-Trailer. (Cnf)
Semi-Finished Inventory	Mfg W	Materials that are no longer in raw-material form, but which have not completed the production cycle to become finished goods. (KA)

Suggest New Words http://www.IDII.com/addword.htm

© 2004 Industrial Data & Information Inc.

Glossary of Supply Chain Terminology

Term	Abbr	Definition
Semi-Trailer	Tr	1. Trailer equipped with rear wheels only. The front of the trailer is supported by landing legs when not hooked to power. 2. Generally used to refer to a full size (45 to 48-foot) trailer; as opposed to a doubles trailer. (Cnf)
Sensitizer	MSDS	A substance which on first exposure causes little or no reaction, but which on repeated exposure may cause a marked response not necessarily limited to the contact site. Skin sensitization is the most common form of sensitization in the industrial setting, although respiratory sensitization to a few chemicals is also known to occur. (HC)
SEQ	B	Sequence. A sequential number. Common abbreviation used in business and in computer programming.
Sequencing	W	The process of organizing items in a load so they will be in the order needed for production. (KA)
Serial Number	W	A unique identification number assigned to a single item. (KA)
Serpentine	W	A picking path that is in a serpentine pattern of bins to inventory from. Some call this a Z picking path.
Service of Process Agents	Tr	The person that an entity engaged in business must make available to everyone for the purpose of serving notice that a legal action has been instituted against that entity. (WA)
Set High	Conv	Vertical spacing that allows the rollers to be mounted above the frame rails. (Fs)
Set Low	Conv	Vertical spacing that allows the roller to be mounted below the top of the frame rails. (Fs)
SEWA	Org	Southeastern Warehouse Association. See http://www.se-warehouseassoc.org/

Suggest New Words http://www.IDII.com/addword.htm

© 2004 Industrial Data & Information Inc.

Glossary of Supply Chain Terminology

SFA	B	Sales Force Automation. Old term now. See CRM.
Shaft	Conv	A bar usually of steel, to support rotating parts or to transmit power. (Fs)
Sheave	Conv	A grooved pulley wheel for carrying a v-belt. (Fs)
Shelf Life	W	The length of time a product can be kept for sale or use before quality considerations make it necessary or desirable to remove it. (KA)
Shelter	W	A cover that protects the space between the door of a rail car or truck and a warehouse from inclement weather. (KA)
Side Channels	Conv	Members that support the rollers on the side of the conveyor. (Fs)
SHIELD	Org Tr	Shippers for International Electronic Logistics Data. This group promotes e-commerce best practices and expedites international trade through automation.
Ship-Age Limit	W	The final date a perishable product can be shipped to a customer. (KA)
ShipTo	W	The name and address of where the shipment will be delivered. Also known as the consignee.
Shipment	Tr W	1. Lot of freight tendered to carrier by one consignee at one place at one time for delivery to one consignee at one place on one bill of lading. 2. Goods/merchandise in one or more containers, pieces, or parcels for transportation from one shipper to single destination. (Cnf)
Shipper	Tr	The party who tenders the goods to the carrier for movement. Can also be known as the consignee. (WA)

Suggest New Words http://www.IDII.com/addword.htm

© 2004 Industrial Data & Information Inc.

Glossary of Supply Chain Terminology

Term	Tag	Definition
Shipper's Certificate	Tr	Form filled out and presented by shipper to outbound carrier at transit point, together with instructions and inbound carrier's freight bill, asking for reshipping privilege and transit rate on commodity previously brought into transit point. (Cnf)
Shipper's Export Declaration	Tr	Form required by Treasury Department and completed by shipper showing value, weight, consignee, destination, etc., of export shipments, as well as Schedule B identification. Abbreviated SED. (Cnf)
Shipper's Load and Count	Tr	Indicates that the contents of a trailer were loaded and counted by the shipper, the trailer was sealed by the shipper, and the carrier did not observe the loading process. Abbreviated SL&C. (Cnf)
Shipping Information	MSDS	The appropriate name(s), hazard class(es), and identification number(s) as determined by the US DOT, International Regulations, and the International Civil Aviation Organization. (HC)
Shipping Pallet	Pa	Pallet designed to be used for a single one-way trip from shipper to receiver; it is then disposed. See Expendable Pallet. (P1)
Shop Floor Control	Mfg W	The process of monitoring and controlling production or warehousing activities to ensure that procedures are followed. (KA)
Short Shipment	Tr	Piece of freight missing from shipment as stipulated by documents on hand. (Cnf)
Shook	Pa	Cut-to-size pallet parts to be assembled into pallets. (P1)
Shook Grade	Pa	The classification of the quality of pallet parts relative to performance characteristics based on size and distribution of defects, independent of wood species. (P1)

Suggest New Words   http://www.IDII.com/addword.htm

© 2004 Industrial Data & Information Inc.

Glossary of Supply Chain Terminology

Term	Category	Definition
Shrink Wrap	Tr W	A plastic wrap used by shippers to secure cartons on a pallet. (Cnf)
Shrinkage	W	Reduction in bulk measurement of inventory. (KA)
Shroud	W	A protective sheet that covers the top and sides of a load, but which permits air to circulate from the bottom. (KA)
SIC	Std	Standard Industry Classification. Coding system to classify different industries including manufacturers, wholesalers, retailers, and service providers.
Side Mounted Drive	Conv	A drive assembly mounted to the side of the conveyor, normally used when minimum elevations are required. (Fs)
Side Tables	Conv	Steel tables attached to either side of conveyor bed to provide working surface close to conveyor. (Fs)
SIG	Org	Special Interest Group. This specialized group has a very specific purpose to accomplish and belongs to a larger organization. See http://sigchi.org/
Single Deep	W	A bin. See also Double Deep.
Single Source	Gov	Purposely award a procurement to a single source. (EM)
Single Sourcing	Pu	The process of using a single supplier for many goods or services. Price competition is reduced, but interdependence should be increased. (KA)
Single-Wing Pallet	Pa	A pallet with the deckboards extending beyond the edges of the stringers or stringer-boards with the bottom deckboards flush (if present). (P1)
Singulation Mode	Conv	Mode where packages are automatically separated while traveling down the conveyor. (Fs)
SIR	Gov	Screening Information Request. A request for proposal & information to meet potential acquisition requirements. See also IFB and SOW.

Suggest New Words   http://www.IDII.com/addword.htm

© 2004 Industrial Data & Information Inc.

Glossary of Supply Chain Terminology

SITC	Std	Standard International Trade Classification.
SITD	B	Still In The Dark. SITD is an acronym used in messaging other online users.
Site Selection Model	W	A program that helps to determine the best location for a distribution center, or other facility. (KA)
Six Sigma	B	Quality management model.
Skatewheel Conveyor	Conv	A type of wheel conveyor making use of series of skatewheels mounted on common shafts or axles, or mounted on parallel spaced bars on individual axles. (Fs)
Skid	Pa	A pallet having no bottom deck. (P1)
SKU	W	Stock Keeping Unit. A product or a set of products referenced by the manufacturer by a unique part number.
SLA	B C	Service Level Agreement. A written contract where servicing company agrees to adhere to provide service(s) that are at or above a specified service level.
Slat Conveyor	Conv	A conveyor that uses steel or wooden slats mounted on roller chain to transport the product. (Fs)
Slave Drive	Conv	A conveyor drive powered from another conveyor instead of having its own prime power source. (Fs)
Slave Pallet	Pa	Pallet, platform or single, thick panel used as a support base for a palletized load in rack-storage facilities or production systems. (P1)
Sleeper	Tr	Tractor with a sleeping compartment in the cab. (Cnf)
Slider Bed	Conv	A stationary surface on which the carrying run of a belt conveyor slides. (Fs)

Suggest New Words http://www.IDII.com/addword.htm

© 2004 Industrial Data & Information Inc.

Glossary of Supply Chain Terminology

Term	Cat	Definition
Slip Sheet	W	Sheet of plastic, cardboard, or fiberboard used rather than a pallet. Forklift attachment is required to handle pallets with slip-sheets.
SLM	B	Service Lifecycle Management (EM)
Slot	W	A position within a storage area reserved for a particular SKU. See also Bin. (KA)
Slug Mode	Conv	Allows all packages to be released simultaneously. (Fs)
SMB	B	Small to Medium Business. The market segment of businesses that are of small to medium size.
SME	B	Small to Medium Enterprises market. Marketing in software companies will plan target certain size of companies as well as specific vertical markets.
SME	Org	Society of Mechanical Engineers. See http://www.asme.org/
SMI	Pu	Supplier Managed Inventories (EM)
SMM	B Mfg	Small and Medium sized Manufacturers. A grouping of manufacturers by annual revenue size.
SMP	C	Symmetrical Multi-Processing
SMS	C	Shipping Manifest System. Computer software that rates freight (parcels, letters, LTL, and TL). Software also produces manifests and shipping labels.
SNA	C	Systems Network Architecture.
Snub Idler	Conv	Any rollers used to increase the arc of contact between a belt and drive pulley. (Fs)
SO	B	Sales Order. A customer's order is known as a sales order.
SOA	B	Services-Oriented Architecture (EM)
SOAP	Std	Simple Object Access Protocol. Data transfer of business data types using XML and SOAP. See also XML, Cobra, OMG, and OASIS.
SOE	B	Statement Of Expectations (EM)

Suggest New Words   http://www.IDII.com/addword.htm

© 2004 Industrial Data & Information Inc.

Glossary of Supply Chain Terminology

Term	Cat	Definition
Soft Allocation	W	Initial allocation of inventory to a specific line item of an order. A "soft" allocation just commits this inventory to the order, but does NOT specifically identify the bin, lot, serial number, or license plate. Soft allocation is done by the ERP system normally. See Hard Allocation.
Soft Nail	Pa	Pallet nail with a MIBANT angle equal to or greater than 47 degrees. (P1)
Software	C	The coding that instructs a computer how to perform specific functions. (KA)
Softwood	Pa	Wood from coniferous or needle-bearing species of trees (not necessarily soft or low density). (P1)
SO/HO	B	Small Office/Home Office
SOLE	Org	International Society of Logistics. Society of practitioners representing commercial, government, and defense. See http://www.sole.org
Sole Source	Gov	A non-competitive procurement.
Solid Deck Pallet	Pa	A pallet constructed with no spacing between deckboards. (P1)
S&OP	B	Sales and Operations Planning.
SOP	B	Standard Operating Procedures. Documented procedures, found in wide variety of topics, including GMP.
Sortation	W	The process of separating packages according to their destination. (KA)
Sortation Container	Conv	A conveyor that is able to sort different packages or products to specific take-away lines. Abbreviated SC. (Fs)
SOW	Gov	Statement of Work. Detailed description of what the Government desires in services and/or products. See IFB & SIR.
SPA	Gov	Simplified Purchasing Agreement.

Suggest New Words http://www.IDII.com/addword.htm

© 2004 Industrial Data & Information Inc.

Glossary of Supply Chain Terminology

Term	Cat	Definition
Spacer	Pa	A pallet component that is located between top and bottom deckboards or beneath a single top deck. A spacer creates the opening that enables forks to be inserted into the pallet. (KA)
Span	Pa	The distance between stringer or block supports. (P1)
SPC	Mfg	Statistical Process Control.
SPEC	B	Specification. Common abbreviation used in business and in computer programming.
Specific Gravity	MSDS	The weight of a material compared to the weight of an equal volume of water is an expression of the density (or heaviness) of a material. Insoluble materials with specific gravity of less than 1.0 will float in or on water. Insoluble materials with specific gravity greater than 1.0 will sink in water. Most (but not all) flammable liquids have specific gravity less than 1.0 and, if not soluble, will float on water - an important consideration for fire suppression. (HC)
Speed Reducer	Conv	A power transmission mechanism designed to provide a speed for the driven equipment less than that of the prime mover. They are generally totally enclosed to retain lubricant and prevent the entry of foreign material. (Fs)
Special Drawing Rights	B	A concept of payment used in international treaties to determine a payment regime. In essence, SDR's are an international currency derived from the average of certain nation's currencies. Abbreviated SDR. (WA)

Suggest New Words   http://www.IDII.com/addword.htm

© 2004 Industrial Data & Information Inc.

Glossary of Supply Chain Terminology

Term	Type	Definition
Split-Month Billing	W	A method of public warehouse billing for storage in which the customer is billed for all inventory in the warehouse at the beginning of the month, as well as for each unit received during that month. Merchandise received during the first half of the month is billed at a full-month storage rate, while merchandise received after the 15th day of the month is billed at a half-month storage rate. See Anniversary Billing. (KA)
Split Shipment	W	A partial shipment that occurs when a warehouse is unable to fill an entire order. The remainder of the order is backordered. (KA)
Spoilage	W	1) One form of product deterioration. 2) The reduction in an inventory's value resulting from inadequate preservation or excess age. (KA)
Spool Conveyor	Conv	A conveyor where power to the rollers is accomplished by o-rings driven by spools on a rotating shaft. (138-SP, 190-SP, 25l4-SP) (Fs)
Spot Check	W	A method of inspecting a shipment in which only a sampling of the total number of containers or items are received are inspected. (KA)
Spur	Conv	A conveyor section to switch unit loads to and from the mainline. (Fs)
SQL	C Std	Standard Query Language. A common programming language to query and maintain (add, update, and delete) data. SQL varies slightly from database vendor to vendor.
SRM	C	Supplier Relationship Management software module. See also CRM, ERM, and PRM.
SSA	Gov	Source Selection Authority.

Suggest New Words   http://www.IDII.com/addword.htm

© 2004 Industrial Data & Information Inc.

Glossary of Supply Chain Terminology

Term	Type	Definition
SSCC	Std	Serial Shipping Container Code. 18-digit license plate code for a container. UCC & EAN are instrumental in this specification.
SSCF	Org	Sanford Supply Chain Forum. See http://www.stanford.edu/group/scforum/
SSO	Gov	Source Selection Official.
SSWA	Org	Sanitary Suppliers Wholesale Association. See http://www.sswa.com/
Stacked Loads	W	Unit loads on pallets that are placed on top of each other to created a column of unitized loads. (KA)
Stacker	Tr	An individual who loads the freight onto a truck or unloads it. (KA)
Stacking	W	The process of placing merchandise on top of other merchandise. (KA)
Stacking Height	W	The distance as measured from the floor to a point 24 inches or more below the lowest overhead obstruction. Stacking height is usually controlled to maintain clearances required by fire regulations. (KA)
Stacks	W	Refers to product stacked in the warehouse. (KA)
Staging Area	W	Temporary storage in a warehouse or terminal where goods are accumulated adjacent to the dock for final loading. (KA)
Standard Costs	B	The anticipated cost for a product or service. (KA)
Standard International Trade Classification	Std	A numerical code developed by the United Nations and adopted by U.S. airlines as the basis for identifying commodities moving in air freight. (Cnf)
Statue of Limitations	B	The amount of time allotted under statute before an actionable claim expires and the claimant may no longer enforce her right to be heard. (WA)

Suggest New Words http://www.IDII.com/addword.htm

© 2004 Industrial Data & Information Inc.

Glossary of Supply Chain Terminology

Statutory Notice	Gov Tr	Length of time required by law for carriers to give notice of changes in tariffs, rates, rules and regulations — usually 30 days unless otherwise permitted by authority from ICC or other regulatory body. (Cnf)
STB	Org Tr	Surface Transportation Board. See www.stb.dot.gov/
STD	B	Standard. Common abbreviation used in business and in computer programming.
STD	B	Standard Deviation
Stevedore	Tr	Person in charge of loading/unloading ships. (Cnf)
Stevedore Pallet	Pa	A pallet designed for use on seaport shipping docks, normally of heavy-duty, double-wing construction. (P1)
Stevedoring	Tr	The unloading of a vessel when in port. (WA)
Stiff-Stock Steel Nail	Pa	Pallet nail made of medium-high carbon steel without heat treatment and tempering with MIBANT angle between 29 and 46 degrees. (P1)
Stock	W	1) Term used to refer to inventory on hand. 2)The activity of replenishing merchandise in storage. (KA)
Stock Keeping Unit	W	Abbreviated SKU. A product or a set of products referenced by the manufacturer by a unique part number.
Stock Locator System	W	A system that allows all storage spots within a warehouse to be identified with an alpha-numeric code, and tracks the items and quantity in each location. (KA)
Stock Order	B	An order to replenish depleted inventory, as opposed to an order to fill a customer order or manufacturing requirement. (KA)
Stockout	W	An event that occurs when one is out of stock on a specific product. Some software solutions record this event for information and/or input to purchasing.

Suggest New Words http://www.IDII.com/addword.htm

© 2004 Industrial Data & Information Inc.

Glossary of Supply Chain Terminology

Stockout Percentage	B	A customer service measurement that shows a measurement of total stockouts to total orders. (KA)
Stock Picker	W	See Order Picker. (KA)
Stock Report	B	A record of items on hand by type and number based on the paper recording of receipts and shipments during a given period. (KA)
Stock Requisition	B	An intracompany form used to authorize the removal of merchandise from its storage location. (KA)
Stock Rotation	W	The process of moving or replacing merchandise to insure freshness and to maximize shelf life. (KA)
Storage Charge	W	A fee for holding goods at rest. (KA)
Storage Costs	W	The sum total of all costs associated with storage, including inventory costs, warehouse costs, administrative costs, deterioration costs, insurance, and taxes. (KA)
Storage Rate	W	The price charged for storage of merchandise, expressed as a cost per unit per month, or as a cost-per-square-foot (or meter) per month. (KA)
Straight Bill of Lading	Tr	Non-negotiable document provides that shipment is to be delivered direct to party whose name is shown as consignee. Carrier does not require its surrender upon delivery except when needed to identify consignee. See Bill of Lading. (Cnf)
Strap Loading	W	The process of loading merchandise onto a pallet and securing it with metal or plastic straps. (KA)
Strapping	W	A metal or plastic band used to hold cases together in a unit. (KA)

Suggest New Words http://www.IDII.com/addword.htm

© 2004 Industrial Data & Information Inc.

Glossary of Supply Chain Terminology

Strap Slot	Pa	Recess or cutout on the upper edge of the stringer or the bottom of the top deckboard to allow tie-down of a unit load to the pallet deck with strapping/banding, also called the banding notch strapping - thin flat bands used to secure load to pallet. (P1)
Stretch Wrap	W	A process and means of applying a sheet of flexible plastic to packages in such a way that they are secured together in a unitized load. (KA)
Stringer	Pa	Continuous, longitudinal, solid or notched beam-component of the pallet used to support deck components, often identified by location as the outside or center stringer. (P1)
Stringerboard	Pa	In block pallets, continuous, solid board member extending for the full length of the pallet perpendicular to deckboard members and placed between deckboards and blocks. (P1)
Stringer Chord	Pa	The upper edge of a notched stringer. (KA)
Stringer Foot	Pa	The lower edge of a notched stringer. (KA)
Stripping	Tr	Emptying truck of cargo, and arranging shipments by destination. (Cnf)
Stuffing	Tr	Slang term for loading cargo container. (Cnf)
SU	Mfg	Set up. An abbreviation used to denote set-up times or set-up charges. (KA)
Subject Matter Jurisdiction	Tr	The subject matter jurisdiction deals with the power of the court to adjudicate the matter before it and cannot be conferred by agreement. (WA)

Suggest New Words   http://www.IDII.com/addword.htm

© 2004 Industrial Data & Information Inc.

Term	Type	Definition
Subrogation	Tr	The substitution of a person or entity (usually an insurance company) to the rights of another person or entity (usually the insured) regarding a claim or a debt which the former person paid for the later. (WA)
Subrogee	Tr	The person or entity that succeeds the rights of another through subrogation. (WA)
SUM	B	Summary. Common abbreviation used in business and in computer programming.
Sunset	B	To legislate an act out of existence. (WA)
Supply Chain Management	B	Supply Chain Management encompasses the planning and management of all activities involved in sourcing and procurement, conversion, and all Logistics Management activities. Importantly, it also includes coordination and collaboration with channel partners, which can be suppliers, intermediaries, third-party service providers, and customers. In essence, Supply Chain Management integrates supply and demand management within and across companies. Abbreviated SCM. See also SCE and SCP. (Cnf)
Support	Conv	Arrangement of members used to maintain the elevation or alignment of the conveyors. Supports can take the form of hangers, floor supports, or brackets and can be either stationary or portable. (Fs)
SVS	Mfg	Sequence Verification System. Sequencing components and assemblies for the Manufacturing process. A system to control the inventory and sequencing of parts to be utilized.
SWA	Org	Southern Wholesalers Association. See http://www.asaonline.org/

Suggest New Words http://www.IDII.com/addword.htm

© 2004 Industrial Data & Information Inc.

Glossary of Supply Chain Terminology

Switch	Conv	(1) Any device for connecting two or more contiguous package conveyor lines; (2) An electrical control device. (Fs)
System	C	An overall set of computer software, hardware, business processes, and people. I.E., the "ERP system" means the ERP software, plus supporting computer hardware, processes utilized in the business with the ERP software, and people that regularly support or utilize this software.
System 34	C	IBM Computer Series. Came out in the 1970's, replaced by the System 36 & 38.
System 36	C	IBM Computer Series. Came out in the 1970's, replaced by the System 38.
System 38	C	IBM Computer Series. Came out in the 1980's, replaced by the AS/400 and the ISeries.

Suggest New Words http://www.IDII.com/addword.htm

© 2004 Industrial Data & Information Inc.

- T -

TAC	Org	Transportation Association of Canada. See http://www.tac-atc.ca/
Tag	W	A method of identifying an item or shipment. (KA)
Tail End	Conv	Usually the end of a conveyor nearest loading point. (Fs)
Tail Pulley	Conv	A pulley mounted at the tail end of a conveyor, its purpose is to return the belt. (Fs)
Take-it-or-leave-it Pallet	Pa	A pallet fitted with fixed cleats on the top deckboards to permit fork truck tines to pass beneath the unit load and remove it from the pallet. (P1)
Take-Up	Conv	The assembly of the necessary structural and mechanical parts which provide the means to adjust the length of belt and chain to compensate for stretch, shrinkage or wear and to maintain proper tension. (Fs)
Tally	W	A sheet made up when goods are received to count and record their condition on arrival. (KA)
Tangent	Conv	Straight portion after a curve conveyor. (Fs)
Tanktainer	Tr	Tank built into standard container frame and used to transport liquids. (Cnf)
Tapered Roller	Conv	A conical conveyor roller for use in a curve with end and intermediate diameter proportional to their distance from center of curve. (Fs)
Tapered Roller Curve	Conv	A curved section of roller conveyor having tapered rollers. (Fs)
Tare	W	The weight of packaging or containers. Tare weight plus net weight equals gross weight. (KA)

Suggest New Words http://www.IDII.com/addword.htm

© 2004 Industrial Data & Information Inc.

Glossary of Supply Chain Terminology

Term	Cat	Definition
Tare Weight	Tr	1. Weight of container and material used for packing. 2. In transportation terms, it is the weight of the car/truck, exclusive of contents. (Cnf)
Tariff	Tr	A publication that contains a carrier's rates, accessorial charges, and rules. (WA)
T&B	Org	Tibbett & Britten Group plc. See www.tibbett-britten.com/
TBD	B	To Be Determined.
TBG	Org	Tibbett & Britten International. See www.tibbett-britten.com/
TCA	Org	Truckload Carriers Association. See http://www.truckload.org/
TCO	B	Total Cost of Ownership. Look at the total cost of ownership for any new project. I.E., new ERP software – the TOC would include hardware, software, professional services, internal staff costs, software installation, and one to three year of maintenance, upgrade, & enhancement costs.
TCPC	Org	Transportation Consumer Protection Council. http://www.transportlaw.com/
TCP-IP	C	Transmission Control Protocol / Internet Protocol.
TDPU	Tr	Time Definite Pick/Up (EM)
T&E	Tr	Transportation & Exportation
Tender	Tr	An offer of goods for transportation by a shipper, or an offer of delivery by a carrier. (Cnf)
Teratogen	MSDS	Any substance that causes growth abnormalities in embryos, genetic modifications in cells, etc. (HC)
Terminals	W	The term for warehouses in early transportation systems. These storage facilities were at the terminal points for land and sea sport. (KA)

Suggest New Words http://www.IDII.com/addword.htm

© 2004 Industrial Data & Information Inc.

Glossary of Supply Chain Terminology

Term	Abbr	Definition
Threshold Limit Values	MSDS	Expresses the airborne concentration of a material to which nearly all persons can be exposed, day after day, without adverse effects. TLV's are expressed three ways: (HC)
Through Bill of Lading	Tr	This document covers goods moving from point of origin to final destination, even if transfers are made to different carriers in transit. (Cnf) See Bill of Lading.
Throughput	Conv	The quantity or amount of product moved on a conveyor at a given time. (Fs)
Throughput	W	1) The total number of units arriving at and departing from warehouse divided by two. Used in public warehouse rate making to calculate average movement of product. 2) A measure of the amount of work done by a computer. Throughput is dependent on hardware and software, and is more useful than simple measure of hardware speed. (KA)
Third Party Warehouse	W	A warehouse operated by a 3PL that contains the client's inventory. See also 3PL, Public Warehouse, and Contract Warehouse.
T&I	Mfg	Testing & Inspection (EM)
TIA	Org	Transportation Intermediaries Association. See http://www.tianet.org/
Tie	W	Number of Units per Layer
Tie-Sheets	Pa	Pallet-size pieces of rough cardboard or fiberboard used between tiers to stabilize unitized loads. (KA)
Tiedown	W	A system of securing a unit load to a pallet. (KA)
Tier	W	1) A single layer of boxes or bags forming one layer of a unitized load. 2) A set of storage locations that are the same height. (KA)

Suggest New Words http://www.IDII.com/addword.htm

© 2004 Industrial Data & Information Inc.

Glossary of Supply Chain Terminology

Term	Cat	Definition
Time Measurement Unit	Uom	Abbreviated TMU. A unit of measure that is used to simplify and standardize small units of time. A TMU is 0.00001 hour (36 thousandths second). TMUs are used in the standard construction sheets to describe the anticipated time required for specific warehousing functions. (KA)
Time Weighted Average Exposure	MSDS	The airborne concentration of a material to which a person is exposed, averaged over the total exposure time, generally the total workday (8 to 12 hours). Abbreviated as TWA. (HC)
TIN	B	Taxpayer Identification Number. US government requires a TIN on most business transactions for tax reporting purposes.
Tine	W	The horizontal load-lifting portion of a fork on a fork truck. It is the portion of a fork that contacts the load. (KA)
TIRRA	Gov	Trucking Industry Regulatory Reform Act.
Title	Tr	Document that confers on holder right of ownership/possession/transfer of merchandise specified, e.g., bills of lading and warehouse receipts. (Cnf)
TL	Tr	See Truck Load.
TLI	Org	The Logistics Institute. The Logistics Institute at Georgia Tech (TLI) has pioneering research and education programs in supply chain, transportation, and e-logistics. Good classes & white papers. See http://tli.isye.gatech.edu
TLV	MSDS	See Threshold Limit Values.
TLV-C	MSDS	The ceiling exposure limit is the concentration that should never be exceeded, even instantaneously. (HC)

Suggest New Words http://www.IDII.com/addword.htm

© 2004 Industrial Data & Information Inc.

Glossary of Supply Chain Terminology

Term	Cat	Definition
TLV-STEL	MSDS	The short-term exposure limit or maximum concentration for a continuous 15-minute exposure period (maximum of four such periods per day, with at least 60 minutes between exposure periods) and provided the TLV-TWA is not exceeded. (HC)
TLV-TWA	MSDS	The allowable Time Weighted Average concentration for a normal 8-hour workday (40-hour work week). (HC)
T&M	B	Time and Materials.
TMA	Org	Tooling and Manufacturing Association. See http://www.tmanet.com/
TMS	Tr	Transportation Management System. A 'broad' term that includes any type of software dealing with logistics & transportation.
TO	B W	Transfer Order. An internal company order to move inventory from one part of the company to another. This could be stock transfer between warehouses, or warehouse to a van, or division to division.
TOFC	Tr	Trailer On Flat Car. A trailer on a railroad flat car.
Tolerance	B	In statistics, the permitted amount of deviation from the mean or average of a measure. (KA)
Tolerated Accuracy Level	B	The percentage that denotes items counted that were within the recorded book value, plus or minus an acceptable limit. (KA)
Ton Mile	Uom	A measurement that is used to describe efficiency of a carrier. A ton mile is equivalent to one ton of cargo moved one mile. (KA)
Tonnage	Tr	1) The carrying capacity of a ship/vessel. 2) The tax/duty paid on such capacity. 3) The weight a ship will carry, expressed in tons. (Cnf)

Suggest New Words http://www.IDII.com/addword.htm

© 2004 Industrial Data & Information Inc.

Glossary of Supply Chain Terminology

Top Cap	Pa	Panel to be placed on top of a unit load to allow for tight strapping without damaging the unit load. (P1)
Top-Deck of the Pallet	Pa	The assembly of deckboards comprising the upper load-carrying surface of the pallet. (P1)
TOR	W	Top of Roller.
TOT	B	Total. Common abbreviation used in business and in computer programming.
Total Load	Conv	Amount of weight distributed over the entire length of a conveyor. (Fs)
Total Quality Management	B	Abbreviated TQM. An interfunctional approach to process improvement that defines quality by looking at customer expectations. The company then focuses on the prevention, detection, and elimination of defects or other quality problems. (KA)
TowMotor	W	Another name for a forklift, or a vehicle with horizontal lift. A vehicle to move freight within a warehouse or on the dock. AKA Fork-Lift, Hi-Lo or Lift-Truck. See Fork-Lift.

Suggest New Words   http://www.IDII.com/addword.htm

© 2004 Industrial Data & Information Inc.

Glossary of Supply Chain Terminology

Toxic	MSDS	TOXIC: Refers to a chemical falling within any of the following toxic categories: A chemical that has a median lethal dose (LD50) of more than 50 milligrams per kilogram, but not more than 500 milligrams per kilogram of body weight when administered orally to albino rats weighing between 200 and 300 milligrams each. A chemical that has a median lethal dose (LD50) of more than 200 milligrams per kilogram, but not more than 1000 milligrams per kilogram of body weight when administered by continuous contact for 24 hours (or less if death occurs within 24 hours) with the bare skin of albino rabbits weighing between 2 and 3 kilograms each. A chemical that has a median lethal concentration (LC50) in air of more than 200 parts per million, but not more than 2000 parts per million by volume of gas or vapor, or more than two milligrams per liter, but not more than 20 milligrams per liter of mist, fume or dust, when administered by continuous inhalation for one hour (or less if death occurs within one hour) to albino rats weighing between 200 and 300 grams each. (HC)
Toxicity	MSDS	The sum of adverse effects resulting from exposure to a material, generally by the mouth, skin, or respiratory tract. (HC)

Suggest New Words   http://www.IDII.com/addword.htm

© 2004 Industrial Data & Information Inc.

Glossary of Supply Chain Terminology

TP	C		Trading Partner. With EDI or XML, the data is coming from or going to a "trading partner". The trading partner is usually a vendor, a carrier, or a bank.
TP	C		Transaction Processing
TPA	B		Trading Partner Agreement. A formal legal document to authorize data exchange between the company and its customers, vendors, carriers, bank, or other trading partner.
TPC	Org		Transaction Processing Performance Council. See http://www.tpc.org/
TPI	C		Tracks Per Inch
TPS	C		Transactions Per Second
TPS	Tr		Transportation Planning and Scheduling.
TQM	B		Total Quality Management.
TRA	Org		Transportation Research Board. See http://www.nas.edu/trb/
Traceability	W		The ability to track a shipment or an item. Any item with a lot number or serial number should be traceable back to the manufacturer, date and location of assembly. See also Lot Number Traceability. (KA)
Tracer	Tr		A form used to implement tracing and information gathering about a lost or misdirected shipment. (KA)
Trackable	W		The collection of Quantity, Gross Weight, Net Weight, Space, and Pallet.
Tracking	Conv		Steering the belt to hold or maintain a desired path. (Fs)
Traffic	Tr		1) Department/division responsible for obtaining most economic commodity classification and method of transportation materials and products. 2) People and/or property carried by transportation companies. (Cnf)

Suggest New Words http://www.IDII.com/addword.htm

© 2004 Industrial Data & Information Inc.

Glossary of Supply Chain Terminology

Term	Type	Definition
Traffic Cop	Conv	A mechanical or electrical mechanism to prevent collision of objects as they merge from two conveyor lines into a single line. (Fs)
Traffic Management	Tr	The selection of transport modes and of specific carriers within the modes. (KA)
Trailer on FlatCar	Tr	See Piggyback. Shipments moving TOFC receive special rates from tariffs provided for that class of traffic. Abbreviated TOFC. (Cnf)
Tramp	Tr	Vessel that does not operate along definite route on fixed schedule, but calls at any port where cargo is available. (Cnf)
Transfer	Conv	A device or series of devices, usually mounted inside a conveyor section, which uses belts, chains, 0-rings, rollers, or skate-wheels, to move products at right angles to adjacent or parallel conveyor lines. (Fs)
Transfer Price	B	The price used when one division of a corporation transfers goods or services to another division. (KA)
Transponder	Tr	Electronic tag attached to a truck or other vehicle. Information is electronically stored in the tag and when a roadside reader reads the tag, its information becomes available.
Transship	Tr	A term commonly used to denote transfer of goods from one means of transportation to another. The re-handling of goods en route. (Cnf)
Trash Conveyor	Conv	A conveyor, normally a belt conveyor equipped with high side guards, used in transporting empty cardboard boxes and paper trash away from working areas. (TH) (Fs)
Tread Plates	Conv	Diamond top steel filler plates used to till gap between rollers on roller conveyor. (Fs)

Suggest New Words http://www.IDII.com/addword.htm

© 2004 Industrial Data & Information Inc.

Glossary of Supply Chain Terminology

Trip	Pa	Consists of four to six handlings of a pallet. (P1)
Trip Charter	Tr	Hiring vessel to haul cargo for special voyage. (Cnf)
Trip Lease	Tr	An arrangement in which a regulated carrier "leases" or hires an owner / operator to make a single run. (Cnf)
Triple	Tr	A combination of vehicles that has a tractor and three trailers. (Cnf)
Tripod Support	Conv	Three legged stand for small roller and skatewheel conveyor. Usually easily moved or aligned to maintain elevation of the conveyor. (Fs)
Troughed Bed	Conv	A conveyor designed with a deep trough used for carrying broken glass, cans, wood chips, stampings, etc. Also used in recycling operations. (TR, CRB) (Fs)
Troughing Attachments	Conv	Angles used on belt conveyors to cup the edge of the belt. (Fs)
Truck Door	W	The part of the warehouse which accommodates loading and unloading of trucks. It includes an overhead door, and may include a dock leveler, a dock shelter, and a concrete pad for the trailer. AKA Dock Door. (KA)
Truckload	Tr	Abbreviated TL. 1. Quantity of freight that will fill a truck. 2. Quantity of freight weighing the maximum legal amount for a particular type of truck. 3. The quantity of freight necessary to qualify for a truckload rate. (Cnf)
TSA	Tr	Transportation Security Administration. Part of the US Government that is now under the DHS.
TSCM	B	Total Supply Chain Management.

Suggest New Words   http://www.IDII.com/addword.htm

© 2004 Industrial Data & Information Inc.

Glossary of Supply Chain Terminology

Term	Cat	Definition
TTM	B	Time To Market. The time it takes a new product to be designed, manufactured, and delivered to the market place. A key metric for how well a company performs in delivering new products.
TTYL	B	Talk To You Later. TTYL is an acronym used in messaging other online users.
Turnbuckle	Conv	A link with a screw thread at both ends, used for tightening the rod, normally used in cross bracing. (Fs)
Turning Wheel	Conv	Wheel mounted on an adjustable bracket to help insure proper package orientation. (Fs)
Turnover Rate	B	The frequency with which total inventory or a specific class of inventory is completely replaced. It is generally stated as the number of turns per year or per month. (KA)
Turntable	Conv	A horizontal, rotateable conveyor mechanism used for transferring objects between conveyors that are in angular relation to one another. (900, 1800, 360E) (Fs)
Turrett Truck	W	Material handling equipment on which the fork rotate 180 degrees to store or retrieve pallets from either side of the vehicle in a narrow aisle. (KA)
TWA	MSDS	See Time Weight Average exposure.
Two-Pulley Hitch	Conv	A special transition section for moving product from horizontal to incline. (TH) (Fs)
Two-Way Entry Pallet	Pa	A pallet with unnotched solid stringers allowing entry only from the ends. (P1)

Suggest New Words http://www.IDII.com/addword.htm

© 2004 Industrial Data & Information Inc.

- U -

UCC	Org	Uniform Code Council. Founded 1972. A not-for-profit standards development organization. The UCC & the EAN International, administers the EAN UCC, UPC and more. See http://www.uc-council.org
UCC	Std	See Uniform Commercial Code.
UCS	Std	Uniform Container Symbol. See UCC 128.
UDDI	Std	Universal Description, Discovery, and Integration. An online directory to facilitate business-to-business electronic commerce. See http://www.uddi.org for specifications, registering your company, or searching for services or products.
UEL	MSDS	See Upper Explosive Limit.
UFL	MSDS	See Upper Flammable Limit.
UI	C	User Interface. How the user will view and interact with the computer application software. Many user interfaces exist - including old-fashioned "green screen" (character based user interface), GUI, web page, RF handheld screen, RF forklift screen, PDA screen, and voice & 3D vision user interfaces.
UKWA	Org	United Kingdom Warehousing Association. The Association represents third party logistics companies in the United Kingdom. See http://www.ukwa.org.uk
ULD	Tr	Unit Load Device. A container utilized for air freight.
ULD	W	Unitized Loading Device. Freight is shipped in a ULD for loading in and out of aircraft.

Suggest New Words http://www.IDII.com/addword.htm

© 2004 Industrial Data & Information Inc.

Glossary of Supply Chain Terminology

Ullage	Tr	Empty space present when container is not full. (Cnf)
Ultra Vires	Tr	To act without apparent authority. (WA)
UM	B	Unified Messaging. UM services combine voice mail, e-mail and fax messages into one mailbox, allowing a user to retrieve and manage all of their messages from a single point, such as a desktop computer or mobile device.
UN	Org	United Nations. See http://www.un.org/
Undercharge	Tr	Where a carrier or its trustee in bankruptcy files a claim against the shipper for that shipper's practice of paying below the carrier's listed tariff. As a result of this practice, several carriers went out of business and brought undercharge claims in the approximated amount of $32 billion. (WA)
Underside Bed Cover	Conv	Sheet metal used to cover the underneath side of a conveyor. (Fs)
Underside Take-Up	Conv	A take-up section located beneath the bed of a belt conveyor. (Fs)
Undertrussing	Conv	Members forming a rigid framework underneath the conveyor, used for supporting the conveyor. (Fs)
Uniform Straight Bill of Lading	Tr	The Uniform Straight Bill of Lading is the contract of carriage published in the National Motor Freight Classification (NMFC) that governs all motor carriers listed as participants in that tariff. (WA)
Unit Load	Pa	Assembly of goods on a pallet for handling, moving, storing and stacking as a single entity. (P1)
Unit of Measure	W	The degree of detail to which we refer about a Commodity. These are usually expressed as Eaches, Cases, Innerpacks, and Pallets. Also known as UOM.

Suggest New Words http://www.IDII.com/addword.htm

© 2004 Industrial Data & Information Inc.

Glossary of Supply Chain Terminology

Unit per Measure	W	A count expressed as a smaller Unit of Measure per so many larger Units of Measure. For example, There are 24 Cans (Eaches) per 1 Case. Cans are the smaller UOM, since more cans fit into the larger Case.
Unitization	W	1) The consolidation of a number of individual items onto one shipping unit for easier handling. 2) The securing or loading of one or more large items of cargo into a single structure or carton. (KA)
Unitize	W	To consolidate packages into a single unit by banding, binding, or wrapping. (KA)
Universal Commercial Code	Std	Abbreviated UCC. The law used in every state except Louisiana to govern business transactions. (KA)
Universal Product Code	Std	Abbreviated UPC. See UPC-A, UPC-E, UPCC, and the UCC for specifications.
Unix	C	A multi-user operating system running on PC's to large mainframes. Most computer manufacturer's sell & support Unix. See also Linux.
Unreasonable Trade Practice	B	A business practice in violation of the Federal Trade Commission Act. (WA)
Unstable	MSDS	Tending toward decomposition or another state, or as produced or transported, will vigorously polymerize, decompose, condense, or become self-reactive under condition of shocks, pressure, or temperature. (HC)
UOM	W	See Unit of Measure.
UPC	Std	Universal Product Code. See UPC-A, UPC-E, UPCC, and the UCC for specifications.

Suggest New Words   http://www.IDII.com/addword.htm

© 2004 Industrial Data & Information Inc.

Glossary of Supply Chain Terminology

Term	Type	Definition
UPC-A	Std	Barcode with a fixed length of 12 digits for point-of-sale retail product usage. The twelve digits break into: First digit is a product indicator. Next 5 digits is manufacturer id. Next 5 digits is product id number. Last digit is a check digit.
UPC-E	Std	Barcode symbology by EAN/UPC that encodes a UCC-12 identification number in 6 digits.
UPCC	Std	Uniform Product Carton Code.
Upper Explosive Limit, Upper Flammable Limit	MSDS	Upper explosive limit (UEL) or upper flammable limit (UFL) of a vapor or gas; the highest concentration (highest percentage of the substance in air) that will produce a flash of fire when an ignition source (heat, arc, or flame) is present. At higher concentrations, the mixture is too "rich" to burn. See LEL. (HC)
Uprights	W	Vertical numbers used in storage rack. (KA)
UPS	Org	United Parcel Service, AKA UPS. One of the largest parcel carriers in North America and worldwide. See http://www.ups.com.
UR	B	You Are. UR is an acronym used in messaging other online users.
URL	C	Uniform Resource Locator. The address of a web site or a web page. For example, http://www.idii.com is an URL and so is ftp.quarterdeck.com
USB	C	Universal Serial Bus. A single USB port can connect a single device or up to 127 computer devices, such as mice, keyboards, modems, and scanners. USB is popular with making new hardware easy to install via Plug-and-Play, hot plugging, and is available on PC's and Macs.
USC	Gov	United States Code.

Suggest New Words   http://www.IDII.com/addword.htm

© 2004 Industrial Data & Information Inc.

USDA	Org	United States Department of Agriculture. See http://www.usda.gov.
USPO	Org	United States Post Office. See http://www.uspo.gov. The USPO is a "carrier" for letters & parcels.

Glossary of Supply Chain Terminology

- V -		- V -
Vacuum Packaging	W	The process of sealing packages by removing nearly all air. (KA)
VAD	B	Value Added Distributor. A company that sells product AND offers services (e.g. vendor certifications, technology seminars). A VAD should have the resources to help complete an installation or assist a VAR in doing such.
VAL	W	Value Added Logistics. See Value Added Services.
Value Added	W	The contribution made by a step in the distribution process to the functionality, usefulness, or value of a product. (KA)
Value Added Services	W	Often the distribution center is required to perform services for customer orders other than picking and packing. These services are termed Value Added Services and they include (but are not limited to): Ticketing, Kit Assembly, Packaging, Final Finishing, Private Labeling, Pallet Labeling, Shrink Wrapping, and White Boxing.
Valuation, Actual	Tr	The actual value of goods shown on a bill of lading by the shipper when the rate to be applied depends on the value of those goods. (Cnf)
VAN	C	Value Added Networks. Commonly deployed for EDI transmission.
Vapor Density	MSDS	The density of a material's vapor compared to the density of the air. If a vapor density is greater than one, it is more dense than air and will drop to the floor or the lowest point available. If the density is less than one, it is lighter than air and will float upwards like helium. (HC)

Suggest New Words   http://www.IDII.com/addword.htm

© 2004 Industrial Data & Information Inc.

Glossary of Supply Chain Terminology

VAR	C	Value Added Resellers. An older term for resellers & integrators of computer software and/or computer hardware. See ISV.
Variable Costs	B	Costs that vary in proportion to activity levels. (KA)
Variable Speed	Conv	A drive or power transmission mechanism that includes a speed changing device. Standard mechanical variable speed ratios 6:1 A.C.; electrical variable speed ratio 10:1. (Fs)
Variance	B	1) The difference between actual and standard costs, or between budgeted and actual expenditures. 2) In statistics, the measure of dispersion of a distribution. (KA)
VAS	C	See Value Added Services.
VAT	B	Value Added Tax (EM)
VAT	Tr	Value Added Tax.
V Belt	Conv	A belt with a trapezoidal cross section for operation in grooved sheaves permitting wedging contact between the belt sides and groove sides. (Fs)
VDT	C	Video Display Terminal. Older terminals running character based applications (not GUI nor Web browser). E.G., Wyse 60, VT220, VT100.
VEND	B	Vendor. Common abbreviation used in business and in computer programming. Sometimes abbreviated as VNDR.
Vendor	B	1) A term used interchangeably with supplier. 2) The company that provides materials for production or resale. (KA)
Vendor Compliance	Pu	A program in which retailers require vendors to provide detailed information, either on bar coded packing slips or via EDI for inbound receipts. (KA)

Suggest New Words http://www.IDII.com/addword.htm

© 2004 Industrial Data & Information Inc.

Glossary of Supply Chain Terminology

Term	Type	Definition
Venue	Tr	The rule of civil procedure that states the geographic location of where an action can be brought. Not to be confused with subject matter jurisdiction that deals with the question of which court is empowered to hear the action. (WA)
Vertical Carousel	W	See Carousel.
Vertical Clearance	W	The distance between the top of a stack and the bottom of obstacles on the veiling of a facility, such as beams, trusses, and sprinklers. (KA)
Very Narrow Aisle	W	Very Narrow Aisle. A warehouse aisle that is purposed designed to be narrow. Normally this type of aisle has a wire imbedded in the center of floor, so that a forklift can sense the wire and stay precisely centered on the wire as it moves down the aisle.
Vessel Ton	Uom	100 Cubic Feet.
VICS	Std	Voluntary Inter-Industry Commerce on standards. Multiple projects including the CPFR. See http://www.vics.org
VIN	Tr	Vehicle Identification Number.
VMI	B	Vendor Managed Inventory
VNA	W	Very Narrow Aisle.
VNDR	B	Vendor. Common abbreviation used in business and in computer programming. Sometimes abbreviated as VEND.
VOC	MSDS	Volatile Organic Content.
Voice Recognition	C	The ability of a computer to receive, reorganize, and understand voice commands as a means of data collection or process control. (KA)
VoIP	C	Voice over Internet Protocol. A cost saving method to add legacy voice transmission to broadband networks (intranet and internet). See also FoIP.

Suggest New Words   http://www.IDII.com/addword.htm

© 2004 Industrial Data & Information Inc.

Glossary of Supply Chain Terminology

Voucher	B	A document authorizing the disbursement of payment that also signals recognition of a service performed or product purchased. (KA)
Voyage Charter	Tr	Engaging services of cargo ship for specified trip from one port to another at established tonnage rate. (Cnf)
VPN	C Std	Virtual Private Network.
VSM	Mfg	Value Stream Mapping.

Suggest New Words http://www.IDII.com/addword.htm

© 2004 Industrial Data & Information Inc.

- W -		- W -
W3C	Org	World Wide Web Consortium. W3C has over 400 Member organizations and was created in October 1994. Its mission is to develop common protocols (E.G., SOAP, XHTML, CSS, etc.) that promote the WWW's growth and ensure its interoperability. See http://www.w3c.org
Wall Bumpers	W	Concrete-filled pipes 12 to 18 inches tall located to the side of the dock opening to protect adjacent walls from the impact of a misaligned truck trailer. (KA)
Wall-To-Wall Inventory	W	A full physical inventory count which includes everything in the warehouse. (KA)
WAN	C	Wide Area Network. See also Network and LAN.
WANE	Org	Wholesalers Association of the Northeast. See http://www.asaonline.org/
WAP	C	Wireless Application Protocol. Enabling technology to take information from the Internet to be viewed on a cell phone or PDA. See also WML.
Warehouse	W	Place for receiving/storing goods and merchandise for hire. Warehouseman is bound to use ordinary diligence in preserving goods. (Cnf)
Warehouse Activity Report	W	A report that details all activities occurring within the warehouse facility, including merchandise arrivals, loading and unloading times and movements. AKA Activity Report. (KA)

Suggest New Words   http://www.IDII.com/addword.htm

© 2004 Industrial Data & Information Inc.

Glossary of Supply Chain Terminology

Warehouse Management System	W	A management information system that controls warehouse activity, furnishing instructions to warehouse resources to manage operations. These systems typically interface with the Host, and RF devices that collect and disseminate information. Related software that may be included or tightly integrated with a WMS includes a Yard Management System, Parcel Manifesting System AKA Shipping Manifesting System, Slotting Optimization, Load Building AKA Load Optimization, and Transportation Management System.
Warehouse Pallet	Pa	A double-face multiple trip returnable pallet intended for general warehouse use. (P1)
Warehouse Receipt	W	A form that contains information describing the merchandise received into a warehouse. (KA)
Warehouse Receipt	W	A receipt, usually non-negotiable, given for goods placed in a warehouse for storage. The receipt is a legal acknowledgement of responsibility of the care of goods. (KA)
Warehouse Within A Warehouse	W	A concept of taking subsets of the warehouse for a dedicated purpose. Each subset would be a "warehouse within a warehouse". E.G., By dedicating a zone (WWAW) for fast moving A items and another zone (WWAW) for B Items, this would improve picking productivity.
Wastage	Tr	Loss of goods due to handling, decay, leakage, shrinkage, etc. (Cnf)

Suggest New Words http://www.IDII.com/addword.htm

© 2004 Industrial Data & Information Inc.

Glossary of Supply Chain Terminology

Term	Cat	Definition
Waste	Mfg	Residue of material from manufacturing operations that results from mishandling, decay, leakage, shrinkage, etc., and that has no value. Frequently it has a negative value because additional costs and logistical efforts must be incurred for disposal. (KA)
Water Reactive	MSDS	A chemical that reacts with water to release a gas that is either flammable or presents a health hazard. (HC)
Wave	W	A wave is a group of outbound orders to be released into picking together.
Wave Planning	W	Planning a wave of outbound orders by selection criteria. A person that plans the wave is called a planner. A planned wave may be one time or put on a regular schedule.
Waybill	Tr	A document containing the description of goods which are part of a common carrier freight shipment. Shows origin, destination, consignee/consignor, and amount charged. Copies of this document travel with goods and are retained by originating/delivering agents. Used by carriers for internal records and control, especially during transit. It is not a transportation contract. (Cnf)
WC	Mfg W	Work Center.
WCM	Mfg	World Class Manufacturing. The philosophy of being the best manufacturer of a product. It implies the constant improvement to remain an industry leader.
WDA	Org	Wholesale Distributors Association. See http://www.asaonline.org/
Weather Seal	W	A rubber or canvas covering that extends out from a dock face to seal the gap between the dock and the trailer's entrance. (KA)

Suggest New Words http://www.IDII.com/addword.htm

© 2004 Industrial Data & Information Inc.

Glossary of Supply Chain Terminology

WECA	Org	Wireless Ethernet Compatibility Alliance – an organization that serves & promotes the wireless industry. See www.wirelessethernet.org.
Weigh-In-Motion	Tr	Technology that weighs a vehicle while the vehicle is in motion.
Weigh-In-Motion	W	A specialized weight scale that is imbedded in the conveyor line and connected to shipping software.
Weight Bumping	Tr	When a household goods carrier falsifies the weight of the shipment by adding extra weight to the trailer before weighing it on a track scale. (WA)
WEP	C Std	Wired Equivalent Privacy. An encryption standard. For those using wireless 802.11, it is best to use an encryption to keep wireless messages secured, such as WEP, AES, DES, or MAC.
Weigh Station	Tr	Permanent station equipped with scales at which motor vehicles transportation property on public highways are required to stop for checking of gross vehicle and/or axle weights. Many states also use portable scales to comply with their weight limits. Often combined with port of entry facilities. (Cnf)
Weight	Tr	In shipping, weight is qualified further as gross (weight of goods and container), net (weight of goods themselves without any container), and legal (similar to net, determined in such manner as law of particular country/jurisdiction may direct). See Gross Weight, Net Weight. (Cnf)
WERC	Org	Warehousing Education and Research Council. Non-profit organization with educational & research focus on warehousing. See http://www.werc.org
WFC	Pu	Weeks Forward Coverage.
WFFSA	Org	Wholesale Florist & Florist Supplier Association. See http://www.wffsa.org/

Suggest New Words http://www.IDII.com/addword.htm

© 2004 Industrial Data & Information Inc.

Glossary of Supply Chain Terminology

Term	Type	Definition
Wharfage	Tr	A charge assessed against a shipping line for using a wharf, or against freight handlers moving over the pier or dock. (KA)
WHO	Org	World Health Org. See www.who.int/en/
WI-FI	Org	The Wi-Fi Alliance is a nonprofit international association formed in 1999 to certify interoperability of wireless Local Area Network products based on IEEE 802.11 specification. See http://www.wi-fi.org
WIM	W	See Weigh-In-Motion.
Windows 9X	C	Refers to either Windows 95 or Windows 98 operating system by Microsoft. Utilized in 1995 through 1999. Windows 95 and 98 are now obsolete. Was replaced by Windows 2000 and then by Windows XP.
Windows 2000	C	An operating system for PC's by Microsoft. Built on Windows NT technology. Now replaced by Windows XP.
Windows NT	C	An operating system for PC's by Microsoft. Designed for client-server (multi-user) operations. Now replaced by Windows 2000 then Windows XP.
Windows XP	C	An operating system for PC's by Microsoft. Designed for client-server (multi-user) operations.
Wing	Pa	Overhang of deckboard end from outside edge of stringer or stringer. (P1)
WIP	Mfg	See Work In Process.
WM	C	The SAP software company has a software module named WM, which is an abbreviation for Warehouse Management.
WM	W	Warehouse Management. Inventory management & inventory control of product within facilities. See WMS

Suggest New Words   http://www.IDII.com/addword.htm

© 2004 Industrial Data & Information Inc.

Glossary of Supply Chain Terminology

Term	Cat	Definition
WML	C	Wireless Markup Language. WML is a markup language that takes the text portions of Web pages and displays them on PDAs and cell phones. See also WAP.
WMS	W	See Warehouse Management Systems.
WO	B Mfg	Work Order. An order to build finished or semi-finished goods from a bill of materials.
Workers' Compensation	B	Formerly known as workmen's compensation, this is the partial or full replacement of an employee's earnings due to an occupational injury or disease. (KA)
Work In Process	Mfg	Orders may require value-added services such as price tag printing and application. Once picks for an order a completed they are tracked in WIP staging. See also Value Added Services.
WORM	C	Write Once Read Many
WSA	Org	Western Suppliers Association. See http://www.asaonline.org/
WSDL	C Std	Web Services Definition Language. See http://www.w3.org/TR/wsdl for more details.
WSWA	Org	Wine & Spirits Wholesalers Association. See http://www.wswa.org/
WTD	B	Week-To-Date. Common abbreviation used in business and in computer programming.
WW	B	Worldwide
WWAN	C	Wireless Wide Area Network.
WWAW	W	See Warehouse Within A Warehouse.
WWW	C	World Wide Web.
WYSIWYG	C	What You See Is What You Get. Programs that visually display documents and/or images in the same form as when they are printed.

Suggest New Words http://www.IDII.com/addword.htm

© 2004 Industrial Data & Information Inc.

- X -

X12	Std	ANSI EDI standards committee has produced the EDI X12 standards for business documents.
XBRL	Std	Stands for eXtensible Business Reporting Language. An open specification that uses XML-based data tags to describe financial data in business reports & databases. This makes pulling and pushing of data faster and more fluid!
XML	Std	Data format specification called eXtensible Markup Language. Standard groups have defined business transaction standards in XML. See OAG, BizTalk, RosettaNet.
XQL	Std	XML Query Language. Ability to query a database via an XML based query.
XSLT	C	eXtensible Stylesheet Language Transformations.

- Y -

Y2K	B C	Year 2000. Older software kept years in a 2-digit year. Year 2000 compliant software has 4 digit year and leap year issues.
Yard	Tr	Unit of track systems within certain area used for storing cars, loading/unloading freight, and making up trains over which movements not authorized by timetable or train order may be made. Subject to prescribed signals/regulations. (Cnf)
YTD	B	Year To Date. Showing a summary, sub-total, or total value that is for a year to date value.

Suggest New Words http://www.IDII.com/addword.htm

© 2004 Industrial Data & Information Inc.

Glossary of Supply Chain Terminology

- Z -

Zero Defects	Mfg W	A long-range objective that strives for defect-free products. (KA)
Zero Pressure Accumulating Conveyor	Conv	A type of conveyor designed to have zero build-up of pressure between adjacent packages or cartons. (Fs)
Z Picking	W	A picking methodology where 1 picker proceeds down an aisle and picks from both rows and continues though the aisle for additional picks in that aisle. By watching the picker, one would view a Z travel path through the aisle if there were 4 bins to pick from. For other methods of picking, see Serpentine Picking, Zone Picking, Pick-to-Light, Pick-to-Carton, Batch Pick, Wave Planning, and Pick-and-Pass.
Zone	W	A section of the warehouse that has exact characteristics based on material handling types and/or inventory management requirements. The Zone is the highest order in the location definition of: Zone, Aisle, Bay, Level, and Position.
Zone Pick	W	Each worker is to pick in a specific area, or zone, of the warehouse. Many warehouse management systems can be configured to zone picking, where workers are limited to one or a small number of zones.
Zone Picking	W	See Zone Pick.

Suggest New Words http://www.IDII.com/addword.htm

© 2004 Industrial Data & Information Inc.

Glossary of Supply Chain Terminology

Need More Information?

For highly educational -

- *White Papers*
- *Free Newsletter – IDII Software Newsletter*
- *Free Newsletter – Supply Chain Events Newsletter*
- *Book Lists on Supply Chain Topics*
- *Educational Publications*
- *Research*
- *List of Trade Associations*
- *List of Educational Publications*

See our website! Explore http://www.IDII.com now!

Suggest Words, Acronyms, Abbreviations

- Enter at http://www.IDII.com/addword.htm
- E-Mail a list to editor@idii.com
- Fax a list to (918) 464-2221

Thank You !

Philip Obal

Suggest New Words http://www.IDII.com/addword.htm

© 2004 Industrial Data & Information Inc.